SEQUENCE ANALYSIS

IN A NUTSHELL

SEQUENCE ANALYSIS

IN A NUTSHELL

A Guide to Common Tools and Databases

Scott Markel and Darryl León

Beijing • Cambridge • Farnham • Köln • Paris • Sebastopol • Taipei • Tokyo

Sequence Analysis in a Nutshell: A Guide to Common Tools and Databases
by Scott Markel and Darryl León

Printed in the United States of America.

Published by O'Reilly & Associates, Inc., 1005 Gravenstein Highway North, Sebastopol, CA 95472.

O'Reilly & Associates books may be purchased for educational, business, or sales promotional use. Online editions are also available for most titles (*safari.oreilly.com*). For more information, contact our corporate/institutional sales department: 800-998-9938 or *corporate@oreilly.com*.

Editor: Lorrie LeJeune

Production Editor: Philip Dangler

Cover Designer: Ellie Volckhausen

Interior Designer: David Futato

Printing History:

February 2003: First Edition.

ISBN: 0-596-00494-X
[M]

Table of Contents

Part II. Tools

Part III. Appendixes

Preface

Gene sequence data is the most abundant type of data available, and there is a rich array of computational methods and tools that can help analyze patterns within that data. This book brings together the detailed terms, definitions, and command-line options found in the key databases and tools used in sequence analysis. It's meant for use by bioinformaticians in both industry and academia, as well as students. This book is a handy resource and an invaluable reference for anyone who needs to know about the practical aspects and mechanics of sequence analysis.

It's no coincidence that the gene sequences of related species of plants, animals, and microorganisms show complex patterns of similarity to one another. This is one of the most fascinating aspects of the study of evolution. In fact, many molecular biologists are convinced that an understanding of sequence evolution is the first step toward understanding evolution itself. The comparison of gene sequences, or biological sequence analysis, is one of the processes used to understand sequence evolution. It is an important discipline within computational biology and bioinformatics.

If you're new to the field, this book won't teach you how to perform sequence analysis, but it will help you sort out the details of the common tools and data sources used for sequence analysis. If sequence analysis is part of your daily lives (as it is for us), you'll want this easy-to-use book on your desk. We've included many references (especially URLs) for further information on the tools we document, but with this book handy we hope you won't need to use them.

Sequence Analysis Tools and Databases

Many of the software tools used in studying genomes involve sequence analysis, which is one of the many subfields of computational molecular biology. The field of sequence analysis includes pattern and motif searching, sequence comparison, multiple sequence alignment, sequence composition determination, and secondary structure prediction. Because sequence data consists primarily of character strings, it's relatively easy to process the sequence entries in a flat file. Bioinformaticians use a variety of different tools to perform sequence analysis, including:

- Standard Unix tools (e.g., the *grep* family, *sed*, *awk*, and *cut*).
- Publicly available tools (e.g., BLAST, the EMBOSS package).
- Open source libaries (e.g., BioPerl, BioJava, BioPython, BioRuby).
- Custom tools.

Finding these tools is pretty easy, but remembering all the command-line options for your favorites is often more difficult.

Nearly all of these tools were written to manipulate and analyze data stored in databases. Many of the most important biological databases have existed for a decade or more, making them almost ancient in this fast-moving field. The first public release of GenBank (Release 3) was in December 1982. There were 606 sequences containing 680,338 basepairs. Release 132 from October 2002 had 19,808,101 sequences containing 26,525,934,656 basepairs. SWISS-PROT has grown from 3939 protein sequences containing 900,163 amino acids (Release 2.0 in September 1986) to 101,602 protein sequences containing 37,315,215 amino acids (Release 40.0 in October 2001).

Plenty of data is available, and finding it is easy. Downloading it is almost as simple, assuming you've got a broadband Internet connection and plenty of disk space. The hard part is dealing with the plethora of flat file formats and trying to remember what their specific field codes mean. Most of us survive by either having hard copies of README files lying around or remembering exactly where to go look for something we need. The need to remember details about our favorite tools and databases prompted us to gather the information and organize it into this book.

How This Book Is Organized

The book is divided into three fundamental areas: data formats, tools, and biological sequence components.

The data format section contains examples of flat files from key databases, the definitions of the codes or fields used in each database, and the sequence feature types/terms and qualifiers for the nucleotide and protein databases.

While there are many useful publicly and commercially available programs, we limited the tools section to popular public domain programs (e.g., BLAST and ClustalW). We also decided to include the EMBOSS programs. These packages are all excellent examples of sequence tools that allow bioinformaticians to easily use the command line to customize their own analyses and workflows. Each program is described briefly, with one or more examples showing how the program may be invoked. We also include the definitions, descriptions, and/or default parameters for each program's command-line options.

The last section of the book concentrates on information essential to understanding the individual components that make up a biological sequence. The tables in this section include nucleotide and protein codes, genetics codes, and other relevant information. The book is organized as follows:

Part I, *Data Formats*

- Chapter 1, *FASTA Format*, describes the most common sequence data format.
- Chapter 2, *GenBank/EMBL/DDBJ*, describes the flat file format, field definitions, and feature tables used in the three most popular sequence databases.
- Chapter 3, *SWISS-PROT*, describes the flat file format, field definitions, and feature tables used with the SWISS-PROT protein database.
- Chapter 4, *Pfam*, describes the flat file format, field definitions used with Pfam, the database for predicting the function of newly discovered proteins.
- Chapter 5, *PROSITE*, describes the flat file format field definitions used with Prosite, one the many popular databases for sequence profiles, patterns, and motifs.

Part II, *Tools*

- Chapter 6, *Readseq*, describes the supported formats and command-line options of Readseq, a program that reads and writes nucleotide and protein sequences in many useful formats.
- Chapter 7, *BLAST*, includes a list of the command-line options used in the various BLAST (Basic Local Alignment Search Tool) programs. We've also included a brief description of each line command and option.
- Chapter 8, *BLAT*, includes the command-line options for BLAT, the BLAST-Like Alignment Tool.
- Chapter 9, *ClustalW*, includes the command-line options for ClustalW, a multiple sequence alignment program for nucleotide sequences or proteins.
- Chapter 10, *HMMER*, describes the respective options for the HMMER (Hidden Markov Model) suite of programs.
- Chapter 11, *MEME/MAST*, shows examples for using MEME (Multiple EM for Motif Elicitation), a tool for discovering motifs in a group of related DNA or protein sequences, and MAST (Motif Alignment and Search Tool), a tool for searching biological sequence databases for sequences that contain one or more of a group of known motifs. We've also included command-line options for each program.
- Chapter 12, *EMBOSS*, includes sequence, aligment, feature, and report formats for the EMBOSS (European Molecular Biology Open Software Suite) tools. The chapter also includes a description, example, and summary of the command-line arguments of each tool in the suite.

Part III, *Appendixes*

- Appendix A, *Nucleotide and Amino Acid Tables*, includes tables of the single-letter nucleotide and amino acid codes, as well as amino acid side chain data.
- Appendix B, *Genetic Codes*, includes the genetic codes for the most common organisms.

- Appendix C, *Resources*, includes useful URLs, further reading, and references to important journal articles.
- Appendix D, *Future Plans*, contains the authors' proposed contribution to the EMBOSS suite.

Assumptions This Book Makes

We assume that you have some familiarity with sequence analysis and its databases and tools, as well as basic working knowledge of your computer environment. For example, you understand how to install a program locally on your machine, and you know how to use command-line options in your tools and on your operating system.

We also assume that the information for each database or tool will not change significantly from the initial writing of the book.

Conventions Used in This Book

We use the following font conventions in this book:

Italic is used for:

- Unix pathnames, filenames, and program names
- Internet addresses, such as domain names and URLs
- New terms where they are defined

Boldface is used for:

- Names of GUI items: window names, buttons, menu choices, etc.

`Constant Width` is used for:

- Command lines and options that should be typed verbatim
- Names and keywords in Java programs, including method names, variable names, and class names
- XML element tags

How to Contact Us

We have tested and verified the information in this book and in the source code to the best of our ability, but given the number of tools described in this book and the rapid pace of technological change, you may find that features have changed or that we have made mistakes. If so, please notify us by writing to:

O'Reilly & Associates
1005 Gravenstein Highway
Sebastopol, CA 95472
800-998-9938 (in the U.S. or Canada)
707-829-0515 (international or local)
707-829-0104 (fax)

To ask technical questions or comment on the book, send email to:

bookquestions@oreilly.com

We have a web site for this book where you can find errata and other information about this book. You can access this page at:

http://www.oreilly.com/catalog/seqanalysis

For more information about this book and others, see the O'Reilly web site:

http://www.oreilly.com

Acknowledgments

We would like to thank Reinhard Schneider and Friedrich von Bohlen at LION bioscience AG (Europe) and Mark Canales and Rudy Potenzone at LION bioscience Inc. (US) for fostering an environment of scientific and technical innovation. Thanks also to Hartmut Voss, Mike Dickson, and Beth Sump for their encouragement and support as we wrote this book. In addition, we would like to thank the past and present architects, developers, software QA members, technical writers, and our officemates at LION. They all asked good questions and made us better at what we do.

Thanks also to Georg Beckmann at Schering AG and Mark Graves at Berlex Laboratories for their real-world problems and our great discussions about how to solve them. We learned much from you.

A special thanks goes to our technical reviewers, Helge Weissig and Cynthia Gibas. Their insightful comments made us rethink the scope of the book and led us to make it more complete.

And finally, we want to thank Lorrie LeJeune, our editor, for planting the seed for this book and working with us on this fun project over the past few months. She made this whole process seem so painless that we're looking forward to working on another book with her. We also want express our gratitude to Philip Dangler, Todd Mezzulo, and the very professional staff at O'Reilly for turning our manuscript into a real book.

From Darryl

I'd like to thank my students, who inspired me to create a practical bioinformatics resource for use in classes and in research.

Susan S. Taylor at UCSD, Bettina Oelke at UCSC Extension, my best friend, Chris Soliz, and my loving parents and siblings have supported me both professionally and personally. My endless thanks go out to them. I'd also like to thank my coauthor, Scott, who listened to my suggestions and shared the desire to create a great book for our readers. I look forward to future collaborations with him.

Most of all, I want to thank my wife, Alison. As a writer herself, she shared my excitement as I finished this book, and understands how much it means to me. Every day we look forward to our future together.

From Scott

As a Christian I want to start by thanking God for His many blessings, including the opportunity to write this book.

Mick Noordewier gave me my first opportunity in bioinformatics when he offered me a job at the R. W. Johnson Pharmaceutical Research Institute (now Johnson & Johnson Pharmaceutical Research and Development). I learned a lot from Mick about the problems scientists really want to solve.

At NetGenics, Mike Dickson and Manuel Glynias believed in me and provided a wonderful environment in which to mix my scientific and software skills. I'll always be grateful.

I've profited greatly from my involvement with the Object Management Group's (OMG) Life Sciences Research (LSR) Domain Task Force. In particular, I'd like to acknowledge the co-submitters and evaluators of the Biomolecular Sequence Analysis specification from whom I learned so much.

Thanks to my parents, Wayne and Caryl Markel, who have always loved me and encouraged me, and showed me how important learning is.

Thanks also to my coauthor, Darryl, and to Alison for her encouragement. Darryl and I discovered a lot about each other and ourselves while writing this book.

My children—Klaudia, Nathan, and Victor—often remind me that there's more to life than work and writing a book. They continually let me re-experience the world through their eyes. I hope they always keep a portion of their childlike innocence.

And finally, my thanks and appreciation go to my wife Danette, the love of my life. Her encouragement, sound advice, and belief in me are truly amazing. Words can only begin to express what she means to me.

I

Data Formats

Bioinformatics, as we know it today, exists because of the vast number of sequence databases created in the last fifteen years. Many of these databases were constructed by scientists who needed a way to organize and annotate the data being generated by their efficient large-sequencing machines. Because these informative sequence files needed to be read by both computers and humans, most sequence databases were designed to use a flat file format. In this section, we explain the more popular flat file formats (GenBank, EMBL, etc.) and focus on describing, in detail, their sometimes cryptic content. While many sequence formats are available, the flat file format is usually used in sequence analysis. Please note that for easy comparison we have provided the same sequence (cyclin-dependent kinase 2) for each of the flat file examples. To give a complete picture of the chosen databases, we have also summarized information related to the feature terms used in the selected sequence flat files.

1
FASTA Format

The most common sequence format you'll encounter is FASTA. This format is quite simple. The first line of a sequence entry consists of ">", followed by an identifier, which contains no whitespace. This can be followed by whitespace and a comment or description. This first line is referred to as the comment or description line. One or more sequence data lines may follow. The length of the sequence data lines may not be constant. Common line lengths are 60, 70, 72, and 80. For details, see the "References" section at the end of this chapter. Example 1-1 contains a sample FASTA entry.

Example 1-1. Sample FASTA entry

```
>gi|29848|emb|X61622.1|HSCDK2MR H.sapiens CDK2 mRNA
ATGGAGAACTTCCAAAAGGTGGAAAAGATCGGAGAGGGCACGTACGGAGTTGTGTACAAAGCCAGAAACA
AGTTGACGGGAGAGGTGGTGGCGCTTAAGAAAATCCGCCTGGACACTGAGACTGAGGGTGTGCCCAGTAC
TGCCATCCGAGAGATCTCTCTGCTTAAGGAGCTTAACCATCCTAATATTGTCAAGCTGCTGGATGTCATT
CACACAGAAAATAAACTCTACCTGGTTTTTGAATTTCTGCACCAAGATCTCAAGAAATTCATGGATGCCT
CTGCTCTCACTGGCATTCCTCTTCCCCTCATCAAGAGCTATCTGTTCCAGCTGCTCCAGGGCCTAGCTTT
CTGCCATTCTCATCGGGTCCTCCACCGAGACCTTAAACCTCAGAATCTGCTTATTAACACAGAGGGGGCC
ATCAAGCTAGCAGACTTTGGACTAGCCAGAGCTTTTGGAGTCCCTGTTCGTACTTACACCCATGAGGTGG
TGACCCTGTGGTACCGAGCTCCTGAAATCCTCCTGGGCTCGAAATATTATTCCACAGCTGTGGACATCTG
GAGCCTGGGCTGCATCTTTGCTGAGATGGTGACTCGCCGGGCCCTGTTCCCTGGAGATTCTGAGATTGAC
CAGCTCTTCCGGATCTTTCGGACTCTGGGGACCCCAGATGAGGTGGTGTGGCCAGGAGTTACTTCTATGC
CTGATTACAAGCCAAGTTTCCCCAAGTGGGCCCGGCAAGATTTTAGTAAAGTTGTACCTCCCCTGGATGA
AGATGGACGGAGCTTGTTATCGCAAATGCTGCACTACGACCCTAACAAGCGGATTTCGGCCAAGGCAGCC
CTGGCTCACCCTTTCTTCCAGGATGTGACCAAGCCAGTACCCCATCTTCGACTCTGATAGCCTTCTTGAA
GCCCCCGACCCTAATCGGCTCACCCTCTCCTCCAGTGTGGGCTTGACCAGCTTGGCCTTGGGCTATTTGG
ACTCAGGTGGGCCCTCTGAACTTGCCTTAAACACTCACCTTCTAGTCTTAACCAGCCAACTCTGGGAATA
CAGGGGTGAAAGGGGGGAACCAGTGAAAATGAAAGGAAGTTTCAGTATTAGATGCACTTAAGTTAGCCTC
CACCACCCTTTCCCCCTTCTCTTAGTTATTGCTGAAGAGGGTTGGTATAAAAATAATTTTAAAAAAGCCT
TCCTACACGTTAGATTTGCCGTACCAATCTCTGAATGCCCCATAATTATTATTTCCAGTGTTTGGGATGA
CCAGGATCCCAAGCCTCCTGCTGCCACAATGTTTATAAAGGCCAAATGATAGCGGGGGCTAAGTTGGTGC
TTTTGAGAATTAAGTAAAACAAAACCACTGGGAGGAGTCTATTTTAAAGAATTCGGTTAAAAAATAGATC
CAATCAGTTTATACCCTAGTTAGTGTTTTCCTCACCTAATAGGCTGGGAGACTGAAGACTCAGCCCGGGT
GGGGGT
```

Many organizations have specific syntax for the description line and have written their own code for parsing and writing FASTA files. Most open source tools expect only the identifier, and treat the rest of the line as a single description string.

A FASTA file may contain more than one sequence entry. The entries are merely concatentated, with the “>” prefixed lines indicating the start of a new sequence entry.

NCBI’s Sequence Identifier Syntax

The National Center for Biotechnology Information (NCBI) uses the following syntax for its BLAST server. NCBI is part of the National Library of Medicine (NLM) at the National Institutes of Health (NIH). The following (including the table) is NCBI’s description. See *ftp://ftp.ncbi.nih.gov/blast/db/README* for details.

The syntax of sequence header lines used by the NCBI BLAST server depends on the database from which each sequence was obtained. The table below lists the identifiers for the databases from which the sequences were derived.

<table>
<tr><th>Database name</th><th>Identifier syntax</th></tr>
<tr><td>GenBank</td><td>gb|accession|locus</td></tr>
<tr><td>EMBL Data Library</td><td>emb|accession|locus</td></tr>
<tr><td>DDBJ, DNA Database of Japan</td><td>dbj|accession|locus</td></tr>
<tr><td>NBRF PIR</td><td>pir||entry</td></tr>
<tr><td>Protein Research Foundation</td><td>prf||name</td></tr>
<tr><td>SWISS-PROT</td><td>sp|accession|entry name</td></tr>
<tr><td>Brookhaven Protein Data Bank</td><td>pdb|entry|chain</td></tr>
<tr><td>Patents</td><td>pat|country|number</td></tr>
<tr><td>GenInfo Backbone Id</td><td>bbs|number</td></tr>
<tr><td>General database identifier</td><td>gnl|database|identifier</td></tr>
<tr><td>NCBI Reference Sequence</td><td>ref|accession|locus</td></tr>
<tr><td>Local Sequence identifier</td><td>lcl|identifier</td></tr>
</table>

For example, an identifier might be “gb|M73307|AGMA13GT”, where the “gb” tag indicates that the identifier refers to a GenBank sequence, “M73307” is its GenBank ACCESSION, and “AGMA13GT” is the GenBank LOCUS.

“gi” identifiers are being assigned by NCBI for all sequences contained within NCBI’s sequence databases. This identifier provides a uniform and stable naming convention whereby a specific sequence is assigned its unique gi identifier. If a nucleotide or protein sequence changes, however, a new gi identifier is assigned, even if the accession number of the record remains unchanged. Thus, gi identifiers provide a mechanism for identifying the exact sequence that was used or retrieved in a given search.

NCBI's Non-Redundant Database Syntax

You should be aware of one additional syntax that's used by the NCBI for their non-redundant database. Since the whole point of the database is to have sequence entries listed only once, the description line syntax allows for more than one set of identifier and description. The sets are delimited by Ctrl-A characters. Here's what NCBI has to say about this.

> These files are all non-redundant; identical sequences are merged into one entry. To be merged two sequences must have identical lengths and every residue (or basepair) at every position must be the same. The FASTA deflines for the different entries that belong to one sequence are separated by control-A's (^A). In the following example, both entries gi|1469284 and gi|1477453 have the same sequence, in every respect.

```
>gi|1469284 (U05042) afuC gene product [Actinobacillus
pleuropneumoniae]^Agi|1477453 (U04954) afuC gene product [Actinobacillus
pleuropneumoniae]
MNNDFLVLKNITKSFGKATVIDNLDLVIKRGTMVTLLGPSGCGKTTVLRLVAGLENPTSGQIFIDGEDVT
KSSIQNRDICIVFQSYALFPHMSIGDNVGYGLRMQGVSNEERKQRVKEALELVDLAGFADRFVDQISGGQ
QQRVALARALVLKPKVLILDEPLSNLDANLRRSMREKIRELQQRLGITSLYVTHDQTEAFAVSDEVIVMN
KGTIMQKARQKIFIYDRILYSLRNFMGESTICDGNLNQGTVSIGDYRFPLHNAADFSVADGACLVGVRPE
AIRLTATGETSQRCQIKSAVYMGNHWEIVANWNGKDVLINANPDQFDPDATKAFIHFTEQGIFLLNKE
```

References

Pearson, W.R., and D. J. Lipman. 1988. Improved Tools for Biological Sequence Analysis. *Proceedings of teh National Academy of Sciences* 85:2444-2448.

NCBI Sequence Identifier Syntax
ftp://ftp.ncbi.nih.gov/blast/db/README

Non-redundant database
ftp://ftp.ncbi.nih.gov/blast/db/README

2 GenBank/EMBL/DDBJ

GenBank is maintained by the National Center for Biotechnology Information (NCBI). It is joined by the DNA Data Bank of Japan (DDBJ, in Mishima, Japan) and the European Molecular Biology Laboratory (EMBL, in Heidelberg, Germany) nucleotide database from the European Bioinformatics Institute (EBI, in Hinxton, UK) to form the International Nucleotide Sequence Database Collaboration. Although the three repositories have separate sites for data submission, they share sequence data and allow daily downloads of sequence files by the public. We're using GenBank Release 132, EMBL Release 72, and DDBJ Release 51.

Example Flat Files

Sequence flat files are frequently used in many software tools. GenBank, DDBJ, and EMBL each have their own specific flat file format. Flat files from each of these databases are shown in the next several sections, and these examples are used to illustrate the field definitions and the feature table sections for each repository. The sequence from cyclin-dependent kinase-2 (CDK2) is used as the example for all of the sequence flat file entries and the fasta file.

GenBank Example Flat File

Example 2-1 contains a sample sequence entry from GenBank. This entry contains terms from the GenBank Field Definitions and the DDBJ/EMBL/GenBank Feature Table, discussed later in this chapter.

Example 2-1. Sample Genbank entry

```
LOCUS       HSCDK2MR               1476 bp    mRNA    linear   PRI 15-JAN-1992
DEFINITION  H.sapiens CDK2 mRNA.
ACCESSION   X61622
VERSION     X61622.1  GI:29848
```

Example 2-1. Sample Genbank entry (continued)

```
KEYWORDS    CDK2 gene; cell cycle regulation protein; cyclin A binding; protein
            kinase.
SOURCE      Homo sapiens (human)
  ORGANISM  Homo sapiens
            Eukaryota; Metazoa; Chordata; Craniata; Vertebrata; Euteleostomi;
            Mammalia; Eutheria; Primates; Catarrhini; Hominidae; Homo.
REFERENCE   1  (bases 1 to 1476)
  AUTHORS   Elledge,S.J. and Spottswood,M.R.
  TITLE     A new human p34 protein kinase, CDK2, identified by complementation
            of a cdc28 mutation in Saccharomyces cerevisiae, is a homolog of
            Xenopus Eg1
  JOURNAL   EMBO J. 10 (9), 2653-2659 (1991)
  MEDLINE   91330891
REFERENCE   2  (bases 1 to 1476)
  AUTHORS   Elledge,S.J.
  TITLE     Direct Submission
  JOURNAL   Submitted (28-NOV-1991) S.J. Elledge, Dept. of Biochemistry, Baylor
            College of Medicine, 1 Baylor Place, Houston, TX 77030, USA
FEATURES             Location/Qualifiers
     source          1..1476
                     /organism="Homo sapiens"
                     /db_xref="taxon:9606"
                     /clone="pSE1000"
                     /cell_line="EBV transformed Human peripheral lymphocyte
                     (B-cell)"
                     /clone_lib="lambda YES-R cDNA library"
     gene            1..1476
                     /gene="CDK2"
     CDS             1..897
                     /gene="CDK2"
                     /function="protein kinase"
                     /note="cell division kinase. CDC2 homolog"
                     /codon_start=1
                     /protein_id="CAA43807.1"
                     /db_xref="GI:29849"
                     /db_xref="SWISS-PROT:P24941"
                     /translation="MENFQKVEKIGEGTYGVVYKARNKLTGEVVALKKIRLDTETEGV
                     PSTAIREISLLKELNHPNIVKLLDVIHTENKLYLVFEFLHQDLKKFMDASALTGIPLP
                     LIKSYLFQLLQGLAFCHSHRVLHRDLKPQNLLINTEGAIKLADFGLARAFGVPVRTYT
                     HEVVTLWYRAPEILLGSKYYSTAVDIWSLGCIFAEMVTRRALFPGDSEIDQLFRIFRT
                     LGTPDEVVWPGVTSMPDYKPSFPKWARQDFSKVVPPLDEDGRSLLSQMLHYDPNKRIS
                     AKAALAHPFFQDVTKPVPHLRL"
BASE COUNT      368 a    372 c    351 g    385 t
ORIGIN
        1 atggagaact tccaaaaggt ggaaaagatc ggagagggca cgtacggagt tgtgtacaaa
       61 gccagaaaca agttgacggg agaggtggtg gcgcttaaga aaatccgcct ggacactgag
      121 actgagggtg tgcccagtac tgccatccga gagatctctc tgcttaagga gcttaaccat
      181 cctaatattg tcaagctgct ggatgtcatt cacacagaaa ataaactcta cctggttttt
      241 gaatttctgc accaagatct caagaaattc atggatgcct ctgctctcac tggcattcct
      301 cttcccctca tcaagagcta tctgttccag ctgctccagg gcctagcttt ctgccattct
      361 catcgggtcc tccaccgaga ccttaaacct cagaatctgc ttattaacac agagggggcc
      421 atcaagctag cagactttgg actagccaga gcttttggag tccctgttcg tacttacacc
```

Example 2-1. Sample Genbank entry (continued)

```
      481 catgaggtgg tgaccctgtg gtaccgagct cctgaaatcc tcctgggctc gaaatattat
      541 tccacagctg tggacatctg gagcctgggc tgcatctttg ctgagatggt gactcgccgg
      601 gccctgttcc ctggagattc tgagattgac cagctcttcc ggatctttcg gactctgggg
      661 accccagatg aggtggtgtg gccaggagtt acttctatgc ctgattacaa gccaagtttc
      721 cccaagtggg cccggcaaga ttttagtaaa gttgtacctc ccctggatga agatggacgg
      781 agcttgttat cgcaaatgct gcactacgac cctaacaagc ggatttcggc caaggcagcc
      841 ctggctcacc ctttcttcca ggatgtgacc aagccagtac cccatcttcg actctgatag
      901 ccttcttgaa gcccccgacc ctaatcggct caccctctcc tccagtgtgg gcttgaccag
      961 cttggccttg ggctatttgg actcaggtgg gccctctgaa cttgccttaa acactcacct
     1021 tctagtctta accagccaac tctgggaata caggggtgaa aggggggaac cagtgaaaat
     1081 gaaaggaagt ttcagtatta gatgcactta agttagcctc caccaccctt tcccccttct
     1141 cttagttatt gctgaagagg gttggtataa aaataatttt aaaaaagcct tcctacacgt
     1201 tagatttgcc gtaccaatct ctgaatgccc cataattatt atttccagtg tttgggatga
     1261 ccaggatccc aagcctcctg ctgccacaat gtttataaag gccaaatgat agcgggggct
     1321 aagttggtgc ttttgagaat taagtaaaac aaaaccactg ggaggagtct attttaaaga
     1381 attcggttaa aaaatagatc caatcagttt ataccctagt tagtgttttc ctcacctaat
     1441 aggctgggag actgaagact cagcccgggt gggggt
//
```

DDBJ Example Flat File

Example 2-2 contains a sample sequence entry from DDBJ. This entry contains terms from the DDBJ Field Definitions and the DDBJ/EMBL/GenBank Feature Table, discussed later in this chapter.

Example 2-2. Sample DDBJ entry

```
LOCUS       HSCDK2MR                1476 bp    RNA     linear   HUM 15-JAN-1992
DEFINITION  H.sapiens CDK2 mRNA.
ACCESSION   X61622
VERSION     X61622.1
KEYWORDS    CDK2 gene; cell cycle regulation protein; cyclin A binding; protein
            kinase.
SOURCE      human.
  ORGANISM  Homo sapiens
            Eukaryota; Metazoa; Chordata; Craniata; Vertebrata; Euteleostomi;
            Mammalia; Eutheria; Primates; Catarrhini; Hominidae; Homo.
REFERENCE   1  (bases 1 to 1476)
  AUTHORS   Elledge,S.J. and Spottswood,M.R.
  TITLE     A new human p34 protein kinase, CDK2, identified by complementation
            of a cdc28 mutation in Saccharomyces cerevisiae, is a homolog of
            Xenopus Eg1
  JOURNAL   EMBO J. 10, 2653-2659(1991).
  MEDLINE   91330891
REFERENCE   2  (bases 1 to 1476)
  AUTHORS   Elledge,S.J.
  JOURNAL   Submitted (28-NOV-1991) to the EMBL/GenBank/DDBJ databases. S.J.
            Elledge, Dept. of Biochemistry, Baylor College of Medicine, 1 Baylor
            Place, Houston, TX 77030, USA
FEATURES             Location/Qualifiers
     source          1..1476
                     /db_xref="taxon:9606"
```

Example 2-2. Sample DDBJ entry (continued)

```
                     /organism="Homo sapiens"
                     /cell_line="EBV transformed Human peripheral lymphocyte
                     (B-cell)"
                     /clone_lib="lambda YES-R cDNA library"
                     /clone="pSE1000"
     CDS             1..897
                     /db_xref="SWISS-PROT:P24941"
                     /note="cell division kinase. CDC2 homolog"
                     /gene="CDK2"
                     /function="protein kinase"
                     /protein_id="CAA43807.1"
                     /translation="MENFQKVEKIGEGTYGVVYKARNKLTGEVVALKKIRLDTETEGVP
                     STAIREISLLKELNHPNIVKLLDVIHTENKLYLVFEFLHQDLKKFMDASALTGIPLPLI
                     KSYLFQLLQGLAFCHSHRVLHRDLKPQNLLINTEGAIKLADFGLARAFGVPVRTYTHEV
                     VTLWYRAPEILLGSKYYSTAVDIWSLGCIFAEMVTRRALFPGDSEIDQLFRIFRTLGTP
                     DEVVWPGVTSMPDYKPSFPKWARQDFSKVVPPLDEDGRSLLSQMLHYDPNKRISAKAAL
                     AHPFFQDVTKPVPHLRL"
BASE COUNT      368 a    372 c    351 g    385 t
ORIGIN
        1 atggagaact tccaaaaggt ggaaaagatc ggagagggca cgtacggagt tgtgtacaaa
       61 gccagaaaca agttgacggg agaggtggtg gcgcttaaga aaatccgcct ggacactgag
      121 actgagggtg tgcccagtac tgccatccga gagatctctc tgcttaagga gcttaaccat
      181 cctaatattg tcaagctgct ggatgtcatt cacacagaaa ataaactcta cctggttttt
      241 gaatttctgc accaagatct caagaaattc atggatgcct ctgctctcac tggcattcct
      301 cttcccctca tcaagagcta tctgttccag ctgctccagg gcctagcttt ctgccattct
      361 catcgggtcc tccaccgaga ccttaaacct cagaatctgc ttattaacac agaggggggcc
      421 atcaagctag cagactttgg actagccaga gcttttggag tccctgttcg tacttacacc
      481 catgaggtgg tgaccctgtg gtaccgagct cctgaaatcc tcctgggctc gaaatattat
      541 tccacagctg tggacatctg gagcctgggc tgcatctttg ctgagatggt gactcgccgg
      601 gccctgttcc ctggagattc tgagattgac cagctcttcc ggatctttcg gactctgggg
      661 accccagatg aggtggtgtg gccaggagtt acttctatgc ctgattacaa gccaagtttc
      721 cccaagtggg cccggcaaga ttttagtaaa gttgtacctc ccctggatga agatggacgg
      781 agcttgttat cgcaaatgct gcactacgac cctaacaagc ggatttcggc caaggcagcc
      841 ctggctcacc ctttcttcca ggatgtgacc aagccagtac cccatcttcg actctgatag
      901 ccttcttgaa gcccccgacc ctaatcggct caccctctcc tccagtgtgg gcttgaccag
      961 cttggccttg ggctatttgg actcaggtgg gccctctgaa cttgccttaa acactcacct
     1021 tctagtctta accagccaac tctgggaata caggggtgaa agggggggaac cagtgaaaat
     1081 gaaaggaagt ttcagtatta gatgcactta agttagcctc caccacccctt tccccccttct
     1141 cttagttatt gctgaagagg gttggtataa aaataatttt aaaaaagcct tcctacacgt
     1201 tagatttgcc gtaccaatct ctgaatgccc cataattatt atttccagtg tttgggatga
     1261 ccaggatccc aagcctcctg ctgccacaat gtttataaag gccaaatgat agcgggggct
     1321 aagttggtgc ttttgagaat taagtaaaac aaaaccactg ggaggagtct attttaaaga
     1381 attcggttaa aaaatagatc caatcagttt ataccctagt tagtgttttc ctcacctaat
     1441 aggctgggag actgaagact cagcccgggt gggggt
//
```

GenBank/DDBJ Field Definitions

The field terms found in GenBank/DDBJ sequence flat files are used to help organize the information for human readabilty and machine parsing. There are several

GenBank/DDBJ field terms found in a sequence flat file, but the repositories themselves share the same field definitions. Table 2-1 summarizes each of the field definitions.

Table 2-1. GenBank/DDBJ field definitions

Field	Description
LOCUS	A short mnemonic name for the entry, chosen to suggest the sequence's definition. Mandatory keyword/exactly one record.
DEFINITION	A concise description of the sequence. Mandatory keyword/one or more records.
ACCESSION	The primary accession number is a unique, unchanging code assigned to each entry. Mandatory keyword/one or more records.
VERSION	A compound identifier consisting of the primary accession number and a numeric version number associated with the current version of the sequence data in the record. This is followed by an integer key (a "GI") assigned to the sequence by NCBI. Mandatory keyword/exactly one record.
NID	An alternative method of presenting the NCBI GI identifier (described above). The NID is obsolete and was removed from the GenBank flat file format in December 1999.
KEYWORDS	Short phrases describing gene products and other information about an entry. Mandatory keyword in all annotated entries/one or more records.
SEGMENT	Information on the order in which this entry appears in a series of discontinuous sequences from the same molecule. Optional keyword (only in segmented entries)/exactly one record.
SOURCE	Common name of the organism or the name most frequently used in the literature. Mandatory keyword in all annotated entries/one or more records/includes one subkeyword.
ORGANISM	Formal scientific name of the organism (first line) and taxonomic classification levels (second and subsequent lines). Mandatory subkeyword in all annotated entries/two or more records.
REFERENCE	Citations for all articles containing data reported in this entry. Includes four subkeywords and may repeat. Mandatory keyword/one or more records.
AUTHORS	Lists the authors of the citation. Mandatory subkeyword/one or more records.
TITLE	Full title of citation. Optional subkeyword (present in all but unpublished citations)/one or more records.
JOURNAL	Lists the journal name, volume, year, and page numbers of the citation. Mandatory subkeyword/one or more records.
MEDLINE	Provides the Medline unique identifier for a citation. Optional subkeyword/one record.
PUBMED	Provides the PubMed unique identifier for a citation. Optional subkeyword/one record.
REMARK	Specifies the relevance of a citation to an entry. Optional subkeyword/one or more records.
COMMENT	Cross-references to other sequence entries, comparisons to other collections, notes of changes in LOCUS names, and other remarks. Optional keyword/one or more records/may include blank records.
FEATURES	Table containing information on portions of the sequence that code for proteins and RNA molecules and information on experimentally determined sites of biological significance. Optional keyword/one or more records.
BASE COUNT	Summary of the number of occurrences of each base code in the sequence. Mandatory keyword/exactly one record.
ORIGIN	Specification of how the first base of the reported sequence is operationally located within the genome. Where possible, this includes its location within a larger genetic map. Mandatory keyword/exactly one record.
//	Entry termination symbol. Mandatory at the end of an entry/exactly one record.

EMBL Example Flat File

Example 2-3 contains a sample sequence entry from EMBL. This entry contains terms from the EMBL Field Definitions and the DDBJ/EMBL/GenBank Feature Table, discussed later in this chapter.

Example 2-3. Sample EMBL entry

```
ID   HSCDK2MR   standard; RNA; HUM; 1476 BP.
XX
AC   X61622;
XX
SV   X61622.1
XX
DT   15-JAN-1992 (Rel. 30, Created)
DT   15-JAN-1992 (Rel. 30, Last updated, Version 1)
XX
DE   H.sapiens CDK2 mRNA
XX
KW   CDK2 gene; cell cycle regulation protein; cyclin A binding; protein kinase.
XX
OS   Homo sapiens (human)
OC   Eukaryota; Metazoa; Chordata; Craniata; Vertebrata; Euteleostomi; Mammalia;
OC   Eutheria; Primates; Catarrhini; Hominidae; Homo.
XX
RN   [1]
RP   1-1476
RX   MEDLINE; 91330891.
RA   Elledge S.J., Spottswood M.R.;
RT   "A new human p34 protein kinase, CDK2, identified by complementation of a
RT   cdc28 mutation in Saccharomyces cerevisiae, is a homolog of Xenopus Eg1";
RL   EMBO J. 10:2653-2659(1991).
XX
RN   [2]
RP   1-1476
RA   Elledge S.J.;
RT   ;
RL   Submitted (28-NOV-1991) to the EMBL/GenBank/DDBJ databases.
RL   S.J. Elledge, Dept. of Biochemistry, Baylor College of Medicine, 1 Baylor
RL   Place, Houston, TX 77030, USA
XX
DR   GDB; 128984; CDK2.
DR   SWISS-PROT; P24941; CDK2_HUMAN.
XX
FH   Key             Location/Qualifiers
FH
FT   source          1..1476
FT                   /db_xref="taxon:9606"
FT                   /organism="Homo sapiens"
FT                   /cell_line="EBV transformed Human peripheral lymphocyte
FT                   (B-cell)"
FT                   /clone_lib="lambda YES-R cDNA library"
FT                   /clone="pSE1000"
```

Example 2-3. Sample EMBL entry (continued)

```
FT   CDS             1..897
FT                   /db_xref="SWISS-PROT:P24941"
FT                   /note="cell division kinase. CDC2 homolog"
FT                   /gene="CDK2"
FT                   /function="protein kinase"
FT                   /protein_id="CAA43807.1"
FT                   /translation="MENFQKVEKIGEGTYGVVYKARNKLTGEVVALKKIRLDTETEGVP
FT                   STAIREISLLKELNHPNIVKLLDVIHTENKLYLVFEFLHQDLKKFMDASALTGIPLPLI
FT                   KSYLFQLLQGLAFCHSHRVLHRDLKPQNLLINTEGAIKLADFGLARAFGVPVRTYTHEV
FT                   VTLWYRAPEILLGSKYYSTAVDIWSLGCIFAEMVTRRALFPGDSEIDQLFRIFRTLGTP
FT                   DEVVWPGVTSMPDYKPSFPKWARQDFSKVVPPLDEDGRSLLSQMLHYDPNKRISAKAAL
FT                   AHPFFQDVTKPVPHLRL"
XX
SQ   Sequence 1476 BP; 368 A; 372 C; 351 G; 385 T; 0 other;
     atggagaact tccaaaaggt ggaaaagatc ggagagggca cgtacggagt tgtgtacaaa        60
     gccagaaaca agttgacggg agaggtggtg gcgcttaaga aaatccgcct ggacactgag       120
     actgagggtg tgcccagtac tgccatccga gagatctctc tgcttaagga gcttaaccat       180
     cctaatattg tcaagctgct ggatgtcatt cacacagaaa ataaactcta cctggttttt       240
     gaatttctgc accaagatct caagaaattc atggatgcct ctgctctcac tggcattcct       300
     cttcccctca tcaagagcta tctgttccag ctgctccagg gcctagcttt ctgccattct       360
     catcgggtcc tccaccgaga ccttaaacct cagaatctgc ttattaacac agagggggcc       420
     atcaagctag cagactttgg actagccaga gcttttggag tccctgttcg tacttacacc       480
     catgaggtgg tgaccctgtg gtaccgagct cctgaaatcc tcctgggctc gaaatattat       540
     tccacagctg tggacatctg gagcctgggc tgcatctttg ctgagatggt gactcgccgg       600
     gccctgttcc ctggagattc tgagattgac cagctcttcc ggatctttcg gactctgggg       660
     accccagatg aggtggtgtg gccaggagtt acttctatgc ctgattacaa gccaagtttc       720
     cccaagtggg cccggcaaga ttttagtaaa gttgtacctc ccctggatga agatggacgg       780
     agcttgttat cgcaaatgct gcactacgac cctaacaagc ggatttcggc caaggcagcc       840
     ctggctcacc ctttcttcca ggatgtgacc aagccagtac cccatcttcg actctgatag       900
     ccttcttgaa gcccccgacc ctaatcggct caccctctcc tccagtgtgg gcttgaccag       960
     cttggccttg ggctatttgg actcaggtgg gccctctgaa cttgccttaa acactcacct      1020
     tctagtctta accagccaac tctgggaata caggggtgaa agggggggaac cagtgaaaat      1080
     gaaaggaagt ttcagtatta gatgcactta agttagcctc caccacccctt tccccttct      1140
     cttagttatt gctgaagagg gttggtataa aaataatttt aaaaaagcct tcctacacgt      1200
     tagatttgcc gtaccaatct ctgaatgccc cataattatt atttccagtg tttgggatga      1260
     ccaggatccc aagcctcctg ctgccacaat gtttataaag gccaaatgat agcgggggct      1320
     aagttggtgc ttttgagaat taagtaaaac aaaaccactg ggaggagtct attttaaaga      1380
     attcggttaa aaaatagatc caatcagttt ataccctagt tagtgttttc ctcacctaat      1440
     aggctgggag actgaagact cagcccgggt gggggt                                1476
//
```

EMBL Field Definitions

The field codes found in EMBL sequence flat files are used to help organize the information for human readability and machine-based parsing. There are several field codes found in an EMBL sequence flat file, and they are designated with a two-letter abbreviation. Table 2-2 summarizes the content of each field code.

Table 2-2. EMBL field definitions

Line code	Content
ID	Identification
AC	Accession number(s)
SV	New sequence identifier
DT	Date
DE	Description
KW	Keyword
OS	Organism species
OC	Organism classification
OG	Organelle
RN	Reference number
RC	Reference comment(s)
RP	Reference positions
RX	Reference cross-reference(s)
RA	Reference authors
RT	Reference title
RL	Reference location
DR	Database cross-references
FH	Feature table header
FT	Feature table data
CC	Comments or notes
XX	Spacer line
SQ	Sequence header
	(Blanks) Sequence data
//	Termination line

DDBJ/EMBL/GenBank Feature Table

In February 1986, GenBank and EMBL (joined by DDBJ in 1987) started a collaborative effort to create a common feature table format. The overall objective of the feature table was to supply an in-depth vocabulary for describing nucleotide (and protein) features. We're using Version 4 of the feature table.

Features

A feature is a single word or abbreviation indicating a functional role or region associated with a sequence. A list of DDBJ/EMBL/GenBank features is presented in Table 2-3. In the Definition column of the table, the appropriate qualifiers for each feature are in brackets. Mandatory qualifiers are highlighted in bold.

Table 2-3. DDBJ/EMBL/GenBank feature key table

Feature Key	Definition
attenuator	1) region of DNA at which regulation of termination of transcription occurs, which controls the expression of some bacterial operons. 2) sequence segment located between the promoter and the first structural gene that causes partial termination of transcription. [citation, db_xref, evidence, gene, label, map, note, phenotype, usedin]
C_region	Constant region of immunoglobulin light and heavy chains, and T-cell receptor alpha, beta, and gamma chains; includes one or more exons depending on the particular chain. [citation, db_xref, evidence, gene, label, map, note, product, pseudo, standard_name, usedin]
CAAT_signal	CAAT box; part of a conserved sequence located about 75 bp up-stream of the start point of eukaryotic transcription units which may be involved in RNA polymerase binding; consensus=GG (C or T) CAATCT. [citation, db_xref, evidence, gene, label, map, note, usedin]
CDS	Coding sequence; sequence of nucleotides that corresponds with the sequence of amino acids in a protein (location includes stop codon); feature includes amino acid conceptual translation. [allele, citation, codon, codon_start, db_xref, EC_number, evidence, exception, function, gene, label, map, note, number, product, protein_id, pseudo, standard_name, translation, transl_except, transl_table, usedin]
conflict	Independent determinations of the "same" sequence differ at this site or region. [**citation**, db_xref, evidence, label, map, note, gene, replace, usedin]
D-loop	Displacement loop; a region within mitochondrial DNA in which a short stretch of RNA is paired with one strand of DNA, displacing the original partner DNA strand in this region. Also used to describe the displacement of a region of one strand of duplex DNA by a single stranded invader in the reaction catalyzed by RecA protein. [citation, db_xref, evidence, gene, label, map, note, usedin]
D_segment	Diversity segment of immunoglobulin heavy chain, and T-cell receptor beta chain [citation, db_xref, evidence, gene, label, map, note, product, pseudo, standard_name, usedin]
enhancer	A cis-acting sequence that increases the utilization of (some) eukaryotic promoters, and can function in either orientation and in any location (upstream or downstream) relative to the promoter. [citation, db_xref, evidence, gene, label, map, note, standard_name, usedin]
exon	Region of genome that codes for portion of spliced mRNA, rRNA and tRNA; may contain 5' UTR, all CDSs, and 3' UTR. [allele, citation, db_xref, EC_number, evidence, function, gene, label, map, note, number, product, pseudo, standard_name, usedin]
GC_signal	GC box; a conserved GC-rich region located upstream of the start point of eukaryotic transcription units which may occur in multiple copies or in either orientation; consensus=GGGCGG. [citation, db_xref, evidence, gene, label, map, note, usedin]
gene	Region of biological interest identified as a gene and for which a name has been assigned. [allele, citation, db_xref, evidence, function, label, map, note, product, pseudo, phenotype, standard_name, usedin]
iDNA	Intervening DNA; DNA which is eliminated through any of several kinds of recombination. [citation, db_xref, evidence, function, label, gene, map, note, number, standard_name, usedin]
intron	A segment of DNA that is transcribed, but removed from within the transcript by splicing together the sequences (exons) on either side of it. [allele, citation, cons_splice, db_xref, evidence, function, gene, label, map, note, number, standard_name, usedin]
J_segment	Joining segment of immunoglobulin light and heavy chains and T-cell receptor alpha, beta, and gamma chains. [citation, db_xref, evidence, gene, map, note, product, pseudo, standard_name, usedin]

Table 2-3. DDBJ/EMBL/GenBank feature key table (continued)

Feature Key	Definition
LTR	Long terminal repeat, a sequence directly repeated at both ends of a defined sequence, of the sort typically found in retroviruses. [citation, db_xref, evidence, function, gene, label, map, note, standard_name, usedin]
mat_peptide	Mature peptide or protein coding sequence; coding sequence for the mature or final peptide or protein product following post-translational modification; the location does not include the stop codon (unlike the corresponding CDS). [citation, db_xref, EC_number, evidence, function, gene, label, map, note, product, pseudo, standard_name, usedin]
misc_binding	Site in nucleic acid which covalently or non-covalently binds another moiety that cannot be described by any other binding key (primer_bind or protein_bind). [citation, **bound_moiety**, db_xref, evidence, function, gene, label, map, note, usedin]
misc_difference	Feature sequence is different from that presented in the entry and cannot be described by any other Difference key (conflict, unsure, old_sequence, mutation, or modified_base). [citation, clone, db_xref, evidence, gene, label, map, note, phenotype, replace, standard_name, usedin]
misc_feature	Region of biological interest which cannot be described by any other feature key; a new or rare feature. [citation, db_xref, evidence, function, gene, label, map, note, number, phenotype, product, pseudo, standard_name, usedin]
misc_recomb	Site of any generalized, site-specific or replicative recombination event where there is a breakage and reunion of duplex DNA that cannot be described by other recombination keys (iDNA and virion) or qualifiers of source key (/insertion seq, /transposon, /proviral). [citation, db_xref, evidence, gene, label, map, note, **organism**, standard_name, usedin]
misc_RNA	Any transcript or RNA product that cannot be defined by other RNA keys (prim_transcript, precursor_RNA, mRNA, 5' clip, 3' clip, 5' UTR, 3' UTR, exon, CDS, sig_peptide, transit_peptide, mat_peptide, intron, polyA_site, rRNA, tRNA, scRNA, and snRNA). [citation, db_xref, evidence, function, gene, label, map, note, product, standard_name, usedin]
misc_signal	Any region containing a signal controlling or altering gene function or expression that cannot be described by other signal keys (promoter, CAAT_signal, TATA_signal, -35_signal, -10_signal, GC_signal, RBS, polyA_signal, enhancer, attenuator, terminator, and rep_origin). [citation, db_xref, evidence, function, gene, label, map, note, phenotype, standard_name, usedin]
misc_structure	Any secondary or tertiary nucleotide structure or conformation that cannot be described by other Structure keys (stem_loop and D-loop). [citation, db_xref, evidence, function, gene, label, map, note, standard_name, usedin]
modified_base	The indicated nucleotide is a modified nucleotide and should be substituted for by the indicated molecule (given in the mod_base qualifier value). [citation, db_xref, evidence, frequency, gene, label, map, **mod_base**, note, usedin]
mRNA	Messenger RNA; includes 5' untranslated region (5'UTR), coding sequences (CDS, exon) and 3' untranslated region (3'UTR); [allele, citation, db_xref, evidence, function, gene, label, map, note, product, pseudo, standard_name, usedin]
N_region	Extra nucleotides inserted between rearranged immmunoglobulin segments. [citation, db_xref, evidence, gene, label, map, note, product, pseudo, standard_name, usedin]
old_sequence	The presented sequence revises a previous version of the sequence at this location. [**citation**, db_xref, evidence, gene, label, map, note, replace, usedin]
polyA_signal	Recognition region necessary for endonuclease cleavage of an RNA transcript that is followed by polyadenylation; consensus=AATAAA. [citation, db_xref, evidence, gene, label, map, note, usedin]

Table 2-3. DDBJ/EMBL/GenBank feature key table (continued)

Feature Key	Definition
polyA_site	Site on an RNA transcript to which will be added adenine residues by post-transcriptional polyadenylation. [citation, db_xref, evidence, gene, label, map, note, usedin]
precursor_RNA	Any RNA species that is not yet the mature RNA product; may include 5' clipped region (5'clip), 5' untranslated region (5'UTR), coding sequences (CDS, exon), intervening sequences (intron), 3' untranslated region (3'UTR), and 3' clipped region (3'clip). [allele, citation, db_xref, evidence, function, gene, label, map, note, product, standard_name, usedin]
prim_transcript	Primary (initial, unprocessed) transcript; includes 5' clipped region (5'clip), 5' untranslated region (5'UTR), coding sequences (CDS, exon), intervening sequences (intron), 3' untranslated region (3'UTR), and 3' clipped region (3'clip). [allele, citation, db_xref, evidence, function, gene, label, map, note, standard_name, usedin]
primer_bind	Non-covalent primer binding site for initiation of replication, transcription, or reverse transcription; includes site(s) for synthetic e.g., PCR primer elements. [citation, db_xref, evidence, gene, label, map, note, standard_name, PCR_conditions, usedin]
promoter	Region on a DNA molecule involved in RNA polymerase binding to initiate transcription. [citation, db_xref, evidence, gene, function, label, map, note, phenotype, pseudo, standard_name, usedin]
protein_bind	Non-covalent protein binding site on nucleic acid. [**bound_moiety**, citation, db_xref, evidence, function, gene, label, map, note, standard_name, usedin]
RBS	Ribosome binding site. [citation, db_xref, evidence, gene, label, map, note, standard_name, usedin]
repeat_region	Region of genome containing repeating units. [citation, db_xref, evidence, function, gene, insertion_seq, label, map, note, rpt_family, rpt_type, rpt_unit, standard_name, transposon, usedin]
repeat_unit	Single repeat element. [citation, db_xref, evidence, function, gene, label, map, note, rpt_family, rpt_type, rpt_unit, usedin]
rep_origin	Origin of replication; starting site for duplication of nucleic acid to give two identical copies. [citation, db_xref, direction, evidence, gene, label, map, note, standard_name, usedin]
rRNA	Mature ribosomal RNA ; RNA component of the ribonucleoprotein particle (ribosome) which assembles amino acids into proteins. [citation, db_xref, evidence, function, gene, label, map, note, product, pseudo, standard_name, usedin]
S_region	Switch region of immunoglobulin heavy chains; involved in the rearrangement of heavy chain DNA leading to the expression of a different immunoglobulin class from the same B-cell. [citation, db_xref, evidence, gene, label, map, note, product, pseudo, standard_name, usedin]
satellite	Many tandem repeats (identical or related) of a short basic repeating unit; many have a base composition or other property different from the genome average that allows them to be separated from the bulk (main band) genomic DNA. [citation, db_xref, evidence, gene, label, map, note, rpt_type, rpt_family, rpt_unit, standard_name, usedin]
scRNA	Small cytoplasmic RNA; any one of several small cytoplasmic RNA molecules present in the cytoplasm and (sometimes) nucleus of a eukaryote. [citation, db_xref, evidence, function, gene, label, map, note, product, pseudo, standard_name, usedin]

Table 2-3. DDBJ/EMBL/GenBank feature key table (continued)

Feature Key	Definition
sig_peptide	Signal peptide coding sequence; coding sequence for an N-terminal domain of a secreted protein; this domain is involved in attaching nascent polypeptide to the membrane leader sequence. [citation, db_xref, evidence, function, gene, label, map, note, product, pseudo, standard_name, usedin]
snRNA	Small nuclear RNA molecules involved in pre-mRNA splicing and processing. [citation, db_xref, evidence, function, gene, label, map, note, partial, product, pseudo, standard_name, usedin]
snoRNA	Small nucleolar RNA molecules mostly involved in rRNA modification and processing. [citation, db_xref, evidence, function, gene, label, map, note, partial, product, pseudo, standard_name, usedin]
source	Identifies the biological source of the specified span of the sequence; this key is mandatory; more than one source key per sequence is permissable; every entry will have, as a minimum, a single source key spanning the entire sequence or multiple source keys together spanning the entire sequence. [cell_line, cell_type, chromosome, citation, clone, clone_lib, country, cultivar, db_xref, dev_stage, environmental_sample, focus, frequency, germline, haplotype, lab_host, insertion_seq, isolate, isolation_source, label, macronuclear, map, note, organelle, **organism**, plasmid, pop_variant, proviral, rearranged, sequenced_mol, serotype, serovar, sex, specimen_voucher, specific_host, strain, sub_clone, sub_species, sub_strain, tissue_lib, tissue_type, transgenic, transposon, usedin, variety, virion]
stem_loop	Hairpin; a double-helical region formed by base-pairing between adjacent (inverted) complementary sequences in a single strand of RNA or DNA. [citation, db_xref, evidence, function, gene, label, map, note, standard_name, usedin]
STS	Sequence tagged site; short, single-copy DNA sequence that characterizes a mapping landmark on the genome and can be detected by PCR; a region of the genome can be mapped by determining the order of a series of STSs. [citation, db_xref, evidence, gene, label, note, map, standard_name, usedin]
TATA_signal	TATA box; Goldberg-Hogness box; a conserved AT-rich septamer found about 25 bp before the start point of each eukaryotic RNA polymerase II transcript unit which may be involved in positioning the enzyme for correct initiation; consensus=TATA(A or T)A(A or T). [citation, db_xref, evidence, gene, label, map, note, usedin]
terminator	Sequence of DNA located either at the end of the transcript that causes RNA polymerase to terminate transcription. [citation, db_xref, evidence, gene, label, map, note, standard_name, usedin]
transit_peptide	Transit peptide coding sequence; coding sequence for an N-terminal domain of a nuclear-encoded organellar protein; this domain is involved in post-translational import of the protein into the organelle. [citation, db_xref, evidence, function, gene, label, map, note, product, pseudo, standard_name, usedin]
tRNA	Mature transfer RNA, a small RNA molecule (75-85 bases long) that mediates the translation of a nucleic acid sequence into an amino acid sequence. [anticodon, citation, db_xref, evidence, function, gene, label, map, note, product, pseudo, standard_name, usedin]
unsure	Author is unsure of exact sequence in this region. [citation, db_xref, evidence, gene, label, map, note, replace, usedin]
V_region	Variable region of immunoglobulin light and heavy chains, and T-cell receptor alpha, beta, and gamma chains; codes for the variable amino terminal portion; can be composed of V_segments, D_segments, N_regions, and J_segments. [citation, db_xref, evidence, gene, label, map, note, product, pseudo, standard_name, usedin]

Table 2-3. DDBJ/EMBL/GenBank feature key table (continued)

Feature Key	Definition
V_segment	Variable segment of immunoglobulin light and heavy chains, and T-cell receptor alpha, beta, and gamma chains; codes for most of the variable region (V_region) and the last few amino acids of the leader peptide. [citation, db_xref, evidence, gene, label, map, note, product, pseudo, standard_name, usedin]
variation	A related strain contains stable mutations from the same gene (e.g., RFLPs, polymorphisms, etc.) which differ from the presented sequence at this location (and possibly others). [allele, citation, db_xref, evidence, frequency, gene, label, map, note, phenotype, product, replace, standard_name, usedin]
3' clip	3'-most region of a precursor transcript that is clipped off during processing. [allele, citation, db_xref, evidence, function, gene, label, map, note, standard_name, usedin]
3' UTR	Region at the 3' end of a mature transcript (following the stop codon) that is not translated into a protein. [allele, citation, db_xref, evidence, function, gene, label, map, note, standard_name, usedin]
5' clip	5'-most region of a precursor transcript that is clipped off during processing. [allele, citation, db_xref, evidence, function, gene, label, map, note, partial, standard_name, usedin]
5' UTR	Region at the 5' end of a mature transcript (preceding the initiation codon) that is not translated into a protein. [allele, citation, db_xref, evidence, function, gene, label, map, note, partial, standard_name, usedin]
-10_signal	Pribnow box; a conserved region about 10 bp upstream of the start point of bacterial transcription units which may be involved in binding RNA polymerase; consensus=TAtAaT. [citation, db_xref, evidence, gene, label, map, note, standard_name, usedin]
-35_signal	A conserved hexamer about 35 bp upstream of the start point of bacterial transcription units; consensus=TTGACa [ ] or TGTTGACA [ ]; [citation, db_xref, evidence, gene, label, map, note, standard_name, usedin]
-	"-" is a placeholder for no key; should be used when the need is merely to mark region in order to comment on it or to use it in another feature's location. [citation, db_xref, evidence, function, gene, label, map, note, number, phenotype, product, pseudo, standard_name, usedin]

Qualifiers

A qualifer is auxiliary information about a feature. A feature can have one or more qualifiers. However, some features require mandatory qualifers, while others don't need a qualifer at all. Table 2-4 lists all DDBJ/EMBL/GenBank qualifiers.

Table 2-4. DDBJ/EMBL/GenBank qualifier table

/<qualifier>=	Description
/allele=	Name of the allele for the given gene.
/anticodon=	Location of the anticodon of tRNA and the amino acid for which it codes.
/bound_moiety=	Moiety bound.
/cell_line=	Cell line from which the sequence was obtained.
/cell_type=	Cell type from which the sequence was obtained.
/chromosome=	Chromosome (e.g., Chromosome number) from which the sequence was obtained.
/citation=	Reference to a citation listed in the entry reference field.

Table 2-4. DDBJ/EMBL/GenBank qualifier table (continued)

/<qualifier>=	Description
/clone=	Clone from which the sequence was obtained.
/clone_lib=	Clone library from which the sequence was obtained.
/codon=	Specifies a codon which is different from any found in the reference genetic code.
/codon_start=	Indicates the offset at which the first complete codon of a coding feature can be found, relative to the first base of that feature.
/cons_splice=	Differentiates between intron splice sites that conform to the 5'-GT ... AG-3' splice site consensus.
/country=	Country of origin for DNA sample, intended for epidemiological or population studies.
/cultivar=	Cultivar (cultivated variety) of plant from which sequence was obtained.
/db_xref=	Database cross-reference: pointer to related information in another database.
/dev_stage=	If the sequence was obtained from an organism in a specific developmental stage, it is specified with this qualifier.
/direction=	Direction of DNA replication.
/EC_number=	Enzyme Commission number for enzyme product of sequence.
/environmental_sample	Identifies sequences derived by direct molecular isolation (PCR, DGGE, or other anonymous methods) from an environmental sample with no reliable identification of the source organism.
/evidence=	Value indicating the nature of supporting evidence, distinguishing between experimentally determined and theoretically derived data.
/exception=	Indicates that the amino acid or RNA sequence will not translate or agree with the DNA sequence according to standard biological rules
/focus	Defines the source feature of primary biological interest for records that have multiple source features originating from different organisms.
/frequency=	Frequency of the occurrence of a feature.
/function=	Function attributed to a sequence.
/gene=	Symbol of the gene corresponding to a sequence region.
/germline	If the sequence shown is DNA and a member of the immunoglobulin family, this qualifier is used to denote that the sequence is from unrearranged DNA.
/haplotype=	Haplotype of organism from which the sequence was obtained.
/insertion_seq=	Insertion sequence element from which the sequence was obtained.
/isolate=	Individual isolate from which the sequence was obtained.
/isolation_source=	Describes the physical, environmental and/or local geographical source of the biological sample from which the sequence was derived.
/label=	A label used to permanently tag a feature.
/lab_host=	Laboratory host used to propagate the organism from which the sequence was obtained
/map=	Genomic map position of feature.
/macronuclear	If the sequence shown is DNA and from an organism which undergoes chromosomal differentiation between macronuclear and micronuclear stages, this qualifier is used to denote that the sequence is from macronuclear DNA.
/mod_base=	Abbreviation for a modified nucleotide base.
/note=	Any comment or additional information.
/number=	A number to indicate the order of genetic elements (e.g., exons or introns) in the 5' to 3' direction.
/organelle=	Type of membrane-bound intracellular structure from which the sequence was obtained.
/organism=	Scientific name of the organism that provided the sequenced genetic material.

Table 2-4. DDBJ/EMBL/GenBank qualifier table (continued)

/<qualifier>=	Description
/partial	Differentiates between complete and partial regions.
/PCR_conditions=	Description of reaction conditions and components for PCR.
/phenotype=	Phenotype conferred by the feature.
/pop_variant=	Population variant from which the sequence was obtained.
/plasmid=	Name of plasmid from which sequence was obtained.
/product=	Name of a product encoded by a sequence.
/protein_id=	Protein identifier, issued by International collaborators, this qualifier consists of a stable ID portion (3+5 format with 3 position letters and 5 numbers) plus a version number after the decimal point.
/proviral	Denotes that the sequence shown is viral and integrated into another organism's genome.
/pseudo	Indicates that this feature is a non-functional version of the element named by the feature key.
/rearranged	If the sequence shown is DNA and a member of the immunoglobulin family, this qualifier denotes that the sequence is from rearranged DNA.
/replace=	Indicates that the sequence identified a feature's intervals is replaced by the sequence shown in "text".
/rpt_family=	Type of repeated sequence; "Alu" or "Kpn", for example.
/rpt_type=	Organization of repeated sequence.
/rpt_unit=	Identity of repeat unit which constitutes a repeat_region.
/sequenced_mol=	Molecule from which the sequence was obtained.
/serotype=	Serological variety of a species characterized by its antigenic properties.
/serovar=	Serological variety of a species (usually a prokaryote) characterized by its antigenic properties.
/sex=	Sex of the organism from which the sequence was obtained.
/specific_host=	natural host from which the sequence was obtained.
/specimen_voucher=	An identifier of the individual or collection of the source organism and the place where it is currently stored, usually an institution.
/standard_name=	Accepted standard name for this feature.
/strain=	Strain from which sequence was obtained.
/sub_clone=	sub-clone from which sequence was obtained.
/sub_species=	Name of sub-species of organism from which sequence was obtained.
/sub_strain=	sub_strain from which sequence was obtained.
/tissue_lib=	Tissue library from which sequence was obtained.
/tissue_type=	Tissue type from which the sequence was obtained.
/transgenic	Identifies the source feature of the organism which was the recipient of transgenic DNA.
/translation=	Automatically generated one-letter abbreviated amino acid sequence derived from either the universal genetic code or the table as specified in /transl_table and as determined by exceptions in the /transl_except and /codon qualifiers.
/transl_except=	Translational exception: single codon the translation of which does not conform to genetic code defined by Organism and /codon=.
/transl_table=	Definition of genetic code table used if other than universal genetic code table (Tables are described in Appendix B).
/transposon=	Transposable element from which the sequence was obtained.
/usedin=	Indicates that the feature is used in a compound feature in another entry.

Table 2-4. DDBJ/EMBL/GenBank qualifier table (continued)

/<qualifier>=	Description
/variety=	Name of variety (formal Linnean rank) of organism from which the sequence was obtained; use the /cultivar qualifier for cultivated plant varieties.
/virion	Viral genomic sequence as it is encapsidated (distinguished from its proviral form integrated in a host cell's chromosome).

Locations

A location is an instruction for finding a feature in a sequence. A list of DDBJ/EMBL/GenBank locations is presented in Table 2-5.

Table 2-5. DDBJ/EMBL/GenBank location examples

Location	Description
467	Points to a single base in the presented sequence.
340..565	Points to a continuous range of bases bounded by and including the starting and ending bases.
<345..500	Indicates that the exact lower boundary point of a feature is unknown. The location begins at some base previous to the first base specified (which need not be contained in the presented sequence) and continues to and includes the ending base.
<1..888	The feature starts before the first sequenced base and continues to and includes base 888.
(102.110)	Indicates that the exact location is unknown but that it is one of the bases between bases 102 and 110, inclusive.
(23.45)..600	Specifies that the starting point is one of the bases between bases 23 and 45, inclusive, and the end point is base 600.
(122.133)..(204.221)	The feature starts at a base between 122 and 133, inclusive, and ends at a base between 204 and 221, inclusive.
123^124	Points to a site between bases 123 and 124.
145^177	Points to a site between two adjacent bases anywhere between bases 145 and 177.
join(12..78,134..202)	Regions 12 to 78 and 134 to 202 should be joined to form one contiguous sequence.
complement(join(2691..4571,4918..5163))	Joins regions 2691 to 4571 and 4918 to 5163, then complements the joined segments (the feature is on the strand complementary to the presented strand).
join(complement(4918..5163),complement(2691..4571))	Complements regions 4918 to 5163 and 2691 to 4571, then joins the complemented segments (the feature is on the strand complementary to the presented strand).
complement(34..(122.126))	Start at one of the bases complementary to those between 122 and 126 on the presented strand and finish at the base complementary to base 34 (the feature is on the strand complementary to the presented strand).
J00194:100..202	Points to bases 100 to 202, inclusive, in the entry (in this database) with primary accession number "J00194".

References

DDBJ

Tateno, Y., T. Imanishi, S. Miyazaki, K. Fukami-Kobayashi, N. Saitou, H. Sugawara, and T. Gojobori. 2002. DNA Data Bank of Japan (DDBJ) for genome scale research in life science. *Nucleic Acids Research* 30 (1):27–30.

Main site
: *http://www.ddbj.nig.ac.jp/*

Release notes
: *http://www.ddbj.nig.ac.jp/ddbjnew/ddbj_relnote.html*

Download
: *ftp://ftp.ddbj.nig.ac.jp/database/ddbj/*

EMBL

Stoesser, G., W. Baker, A. van den Broek, E. Camon, M. Garcia-Pastor, C. Kanz, T. Kulikova, R. Leinonen, Q. Lin, V. Lombard, R. Lopez, N. Redaschi, P. Stoehr, M. A. Tuli, K. Tzouvara, and R. Vaughan. 2002. The EMBL Nucleotide Sequence Database. *Nucleic Acids Research* 30 (1):21–26.

Main page
: *http://www.ebi.ac.uk/embl/index.html*

Release notes
: *http://www.ebi.ac.uk/embl/Documentation/Release_notes/current/relnotes.html*

User manual
: *http://www.ebi.ac.uk/embl/Documentation/User_manual/usrman.html*

Download
: *ftp://ftp.ebi.ac.uk/pub/databases/embl/*

GenBank

Benson, D.A., I. Karsch-Mizrachi, D. J. Lipman, J. Ostell, B. A. Rapp, and D. L. Wheeler. 2002. GenBank. *Nucleic Acids Research* 30 (1):17–20.

GenBank overview
: *http://www.ncbi.nlm.nih.gov/Genbank/GenbankOverview.html*

Release notes
: *ftp://ftp.ncbi.nih.gov/genbank/gbrel.txt*

Download
: *ftp://ftp.ncbi.nih.gov/genbank/*

DDBJ/EMBL/GenBank Feature Table
: *http://www.ncbi.nlm.nih.gov/projects/collab/FT/index.html*

3
SWISS-PROT

SWISS-PROT is an annotated protein sequence database that was started in 1986. It is currently overseen by the Swiss Institute of Bioinformatics (SIB) in association with the European Bioinformatics Institute (EBI). SWISS-PROT is the preferred protein sequence database for most bioinformaticians because many of the sequence annotations are curated by scientists. TrEMBL, another sequence database, is a computer-annotated supplement that contains all the translations of EMBL nucleotide sequence entries not yet integrated in SWISS-PROT. It has essentially the same sequence flat file format as SWISS-PROT. We're using SWISS-PROT Release 40.

SWISS-PROT Example Flat File

Example 3-1 contains a sequence entry from SWISS-PROT. This entry contains terms from the SWISS-PROT Field Definitions and Feature Table types, discussed later in this chapter.

Example 3-1 . Sample SWISS-PROT sequence entry

```
ID   CDK2_HUMAN     STANDARD;      PRT;   298 AA.
AC   P24941;
DT   01-MAR-1992 (Rel. 21, Created)
DT   01-AUG-1992 (Rel. 23, Last sequence update)
DT   15-JUN-2002 (Rel. 41, Last annotation update)
DE   Cell division protein kinase 2 (EC 2.7.1.-) (p33 protein kinase).
GN   CDK2.
OS   Homo sapiens (Human).
OC   Eukaryota; Metazoa; Chordata; Craniata; Vertebrata; Euteleostomi;
OC   Mammalia; Eutheria; Primates; Catarrhini; Hominidae; Homo.
OX   NCBI_TaxID=9606;
RN   [1]
RP   SEQUENCE FROM N.A.
```

Example 3-1 . Sample SWISS-PROT sequence entry (continued)

```
RX   MEDLINE=91330891; PubMed=1714386;
RA   Elledge S.J., Spottswood M.R.;
RT   "A new human p34 protein kinase, CDK2, identified by complementation
RT   of a cdc28 mutation in Saccharomyces cerevisiae, is a homolog of
RT   Xenopus Eg1.";
RL   EMBO J. 10:2653-2659(1991).
RN   [2]
RP   SEQUENCE FROM N.A.
RX   MEDLINE=91367262; PubMed=1653904;
RA   Tsai L.-H., Harlow E., Meyerson M.;
RT   "Isolation of the human cdk2 gene that encodes the cyclin A- and
RT   adenovirus E1A-associated p33 kinase.";
RL   Nature 353:174-177(1991).
RN   [3]
RP   SEQUENCE FROM N.A.
RX   MEDLINE=92020980; PubMed=1717994;
RA   Ninomiya-Tsuji J., Nomoto S., Yasuda H., Reed S.I., Matsumoto K.;
RT   "Cloning of a human cDNA encoding a CDC2-related kinase by
RT   complementation of a budding yeast cdc28 mutation.";
RL   Proc. Natl. Acad. Sci. U.S.A. 88:9006-9010(1991).
RN   [4]
RP   SEQUENCE FROM N.A.
RC   TISSUE=Placenta;
RA   Strausberg R.;
RL   Submitted (FEB-2001) to the EMBL/GenBank/DDBJ databases.
RN   [5]
RP   PHOSPHORYLATION SITES.
RX   MEDLINE=93010995; PubMed=1396589;
RA   Gu Y., Rosenblatt J., O'Morgan D.O.;
RT   "Cell cycle regulation of CDK2 activity by phosphorylation of Thr160
RT   and Tyr15.";
RL   EMBO J. 11:3995-4005(1992).
RN   [6]
RP   X-RAY CRYSTALLOGRAPHY (2.4 ANGSTROMS).
RX   MEDLINE=93288132; PubMed=8510751;
RA   de Bondt H.L., Rosenblatt J., Jancarik J., Jones H.D.,
RA   Morgan D.O., Kim S.-H.;
RT   "Crystal structure of cyclin-dependent kinase 2.";
RL   Nature 363:595-602(1993).
RN   [7]
RP   X-RAY CRYSTALLOGRAPHY (2.3 ANGSTROMS) OF COMPLEX WITH CYCLIN A.
RX   MEDLINE=95356811; PubMed=7630397;
RA   Jeffrey P.D., Russo A.A., Polyak K., Gibbs E., Hurwitz J.,
RA   Massague J., Pavletich N.P.;
RT   "Mechanism of CDK activation revealed by the structure of a
RT   cyclinA-CDK2 complex.";
RL   Nature 376:313-320(1995).
RN   [8]
RP   X-RAY CRYSTALLOGRAPHY (2.33 ANGSTROMS) OF COMPLEX WITH L868276.
RX   MEDLINE=96181476; PubMed=8610110;
RA   de Azevedo W.F. Jr., Muleer-Dieckmann H.-J., Schulze-Gahmen U.,
RA   Worland P.J., Sausville E., Kim S.-H.;
```

Example 3-1 . Sample SWISS-PROT sequence entry (continued)

```
RT   "Structural basis for specificity and potency of a flavonoid
RT   inhibitor of human CDK2, a cell cycle kinase.";
RL   Proc. Natl. Acad. Sci. U.S.A. 93:2735-2740(1996).
RN   [9]
RP   X-RAY CRYSTALLOGRAPHY (2.3 ANGSTROMS) OF COMPLEX WITH CG2A AND KIP1.
RX   MEDLINE=96300318; PubMed=8684460;
RA   Russo A.A., Jeffrey P.D., Patten A.K., Massague J., Pavletich N.P.;
RT   "Crystal structure of the p27Kip1 cyclin-dependent-kinase inhibitor
RT   bound to the cyclin A-Cdk2 complex.";
RL   Nature 382:325-331(1996).
RN   [10]
RP   X-RAY CRYSTALLOGRAPHY (2.6 ANGSTROMS) OF COMPLEX WITH CG2A.
RX   MEDLINE=96313126; PubMed=8756328;
RA   Russo A.A., Jeffrey P.D., Pavletich N.P.;
RT   "Structural basis of cyclin-dependent kinase activation by
RT   phosphorylation.";
RL   Nat. Struct. Biol. 3:696-700(1996).
RN   [11]
RP   X-RAY CRYSTALLOGRAPHY (1.9 ANGSTROMS).
RX   MEDLINE=97075215; PubMed=8917641;
RA   Schulze-Gahmen U., de Bondt H.L., Kim S.-H.;
RT   "High-resolution crystal structures of human cyclin-dependent kinase
RT   2 with and without ATP: bound waters and natural ligand as guides for
RT   inhibitor design.";
RL   J. Med. Chem. 39:4540-4546(1996).
RN   [12]
RP   X-RAY CRYSTALLOGRAPHY (2.0 ANGSTROMS).
RX   MEDLINE=97475219; PubMed=9334743;
RA   Lawrie A.M., Noble M.E.M., Tunnah P., Brown N.R., Johnson L.N.,
RA   Endicott J.A.;
RT   "Protein kinase inhibition by staurosporine revealed in details of
RT   the molecular interaction with CDK2.";
RL   Nat. Struct. Biol. 4:796-801(1997).
RN   [13]
RP   X-RAY CRYSTALLOGRAPHY (2.6 ANGSTROMS) OF COMPLEX WITG CKS1.
RX   MEDLINE=96182647; PubMed=8601310;
RA   Bourne Y., Watson M.H., Hickey M.J., Holmes W., Rocque W., Reed S.I.,
RA   Tainer J.A.;
RT   "Crystal structure and mutational analysis of the human CDK2 kinase
RT   complex with cell cycle-regulatory protein CksHs1.";
RL   Cell 84:863-874(1996).
RN   [14]
RP   X-RAY CRYSTALLOGRAPHY (2.05 ANGSTROMS).
RX   MEDLINE=98342369; PubMed=9677190;
RA   Gray N.S., Wodicka L., Thunnissen A.-M.W.H., Norman T.C., Kwon S.,
RA   Espinoza F.H., Morgan D.O., Barnes G., Leclerc S., Meijer L.,
RA   Kim S.H., Lockhart D.J., Schultz P.G.;
RT   "Exploiting chemical libraries, structure, and genomics in the search
RT   for kinase inhibitors.";
RL   Science 281:533-538(1998).
CC   -!- FUNCTION: PROBABLY INVOLVED IN THE CONTROL OF THE CELL CYCLE.
CC       INTERACTS WITH CYCLINS A, D, OR E. ACTIVITY OF CDK2 IS MAXIMAL
```

Example 3-1 . Sample SWISS-PROT sequence entry (continued)

```
CC       DURING S PHASE AND G2.
CC   -!- ENZYME REGULATION: PHOSPHORYLATION AT THR-14 OR TYR-15 INACTIVATES
CC       THE ENZYME, WHILE PHOSPHORYLATION AT THR-160 ACTIVATES IT.
CC   -!- SIMILARITY: BELONGS TO THE SER/THR FAMILY OF PROTEIN KINASES.
CC       CDC2/CDKX SUBFAMILY.
CC   --------------------------------------------------------------------------
CC   This SWISS-PROT entry is copyright. It is produced through a collaboration
CC   between  the Swiss Institute of Bioinformatics  and the  EMBL outstation -
CC   the European Bioinformatics Institute.  There are no  restrictions on  its
CC   use  by  non-profit  institutions as long  as its content  is  in  no  way
CC   modified and this statement is not removed.  Usage  by  and for commercial
CC   entities requires a license agreement (See http://www.isb-sib.ch/announce/
CC   or send an email to license@isb-sib.ch).
CC   --------------------------------------------------------------------------
DR   EMBL; X61622; CAA43807.1; -.
DR   EMBL; X62071; CAA43985.1; -.
DR   EMBL; M68520; AAA35667.1; -.
DR   EMBL; BC003065; AAH03065.1; -.
DR   PIR; A41227; A41227.
DR   PIR; S16520; S16520.
DR   PIR; S17873; S17873.
DR   PDB; 1FIN; 27-JAN-97.
DR   PDB; 1HCK; 07-DEC-96.
DR   PDB; 1HCL; 07-DEC-96.
DR   PDB; 1AQ1; 12-NOV-97.
DR   PDB; 1JST; 11-JAN-97.
DR   PDB; 1JSU; 29-JUL-97.
DR   PDB; 1BUH; 09-SEP-98.
DR   PDB; 1B38; 23-DEC-98.
DR   PDB; 1B39; 23-DEC-98.
DR   PDB; 1CKP; 13-JAN-99.
DR   Genew; HGNC:1771; CDK2.
DR   MIM; 116953; -.
DR   InterPro; IPR000719; Euk_pkinase.
DR   InterPro; IPR002290; Ser_thr_pkinase.
DR   Pfam; PF00069; pkinase; 1.
DR   ProDom; PD000001; Euk_pkinase; 1.
DR   SMART; SM00220; S_TKc; 1.
DR   PROSITE; PS00107; PROTEIN_KINASE_ATP; 1.
DR   PROSITE; PS00108; PROTEIN_KINASE_ST; 1.
DR   PROSITE; PS50011; PROTEIN_KINASE_DOM; 1.
KW   Transferase; Serine/threonine-protein kinase; ATP-binding;
KW   Cell cycle; Cell division; Mitosis; Phosphorylation; 3D-structure.
FT   DOMAIN        4    286       PROTEIN KINASE.
FT   NP_BIND      10     18       ATP (BY SIMILARITY).
FT   BINDING      33     33       ATP (BY SIMILARITY).
FT   ACT_SITE    127    127       BY SIMILARITY.
FT   MOD_RES      14     14       PHOSPHORYLATION.
FT   MOD_RES      15     15       PHOSPHORYLATION.
FT   MOD_RES     160    160       PHOSPHORYLATION (BY CAK).
FT   MUTAGEN      14     14       T->A: INCREASE ACTIVITY 2 FOLD.
FT   MUTAGEN      15     15       Y->F: INCREASE ACTIVITY 2 FOLD.
FT   MUTAGEN     160    160       T->A: ABOLISHES ACTIVITY.
```

Example 3-1 . Sample SWISS-PROT sequence entry (continued)

```
FT   TURN          2      3
FT   STRAND        4     12
FT   STRAND       17     23
FT   TURN         24     26
FT   STRAND       29     35
FT   HELIX        46     55
FT   TURN         56     57
FT   TURN         61     62
FT   STRAND       63     63
FT   STRAND       66     72
FT   TURN         73     74
FT   STRAND       75     81
FT   STRAND       85     86
FT   HELIX        87     93
FT   TURN         94     97
FT   HELIX       101    120
FT   TURN        121    122
FT   HELIX       130    132
FT   STRAND      133    135
FT   TURN        137    138
FT   STRAND      141    143
FT   TURN        146    147
FT   HELIX       148    151
FT   STRAND      157    157
FT   TURN        159    160
FT   STRAND      163    163
FT   TURN        167    168
FT   HELIX       171    174
FT   TURN        175    176
FT   TURN        182    182
FT   HELIX       183    198
FT   HELIX       208    219
FT   TURN        224    226
FT   TURN        228    229
FT   HELIX       230    232
FT   TURN        234    235
FT   TURN        238    239
FT   HELIX       248    251
FT   TURN        253    254
FT   HELIX       257    266
FT   TURN        267    267
FT   TURN        271    273
FT   HELIX       277    280
FT   TURN        281    282
FT   HELIX       284    286
FT   TURN        287    288
SQ   SEQUENCE   298 AA;  33929 MW;  F90A0F4E70910B51 CRC64;
     MENFQKVEKI GEGTYGVVYK ARNKLTGEVV ALKKIRLDTE TEGVPSTAIR EISLLKELNH
     PNIVKLLDVI HTENKLYLVF EFLHQDLKKF MDASALTGIP LPLIKSYLFQ LLQGLAFCHS
     HRVLHRDLKP QNLLINTEGA IKLADFGLAR AFGVPVRTYT HEVVTLWYRA PEILLGCKYY
     STAVDIWSLG CIFAEMVTRR ALFPGDSEID QLFRIFRTLG TPDEVVWPGV TSMPDYKPSF
     PKWARQDFSK VVPPLDEDGR SLLSQMLHYD PNKRISAKAA LAHPFFQDVT KPVPHLRL
//
```

SWISS-PROT

SWISS-PROT Field Definitions

The field codes found in a SWISS-PROT (or TrEMBL) sequence flat file are used to help arrange the information for human readabilty and machine-based parsing. There are several SWISS-PROT field codes found in a sequence flat file; they are represented by two-letter abbreviations. Table 3-1 summarizes the contents of each field code.

Table 3-1. SWISS-PROT field defintitions

Line code	Content
ID	Identification
AC	Accession number(s)
DT	Date
DE	Description
GN	Gene name(s)
OS	Organism species
OG	Organelle
OC	Organism classification
OX	Taxonomy cross-reference(s)
RN	Reference number
RP	Reference position
RC	Reference comment(s)
RX	Reference cross-reference(s)
RA	Reference authors
RT	Reference title
RL	Reference location
CC	Comments or notes
DR	Database cross-references
KW	Keywords
FT	Feature table data
SQ	Sequence header
	(blanks) sequence data
//	Termination line

SWISS-PROT Feature Table

A feature is a single word or abbreviation indicating a functional role or region associated with a sequence. A list of SWISS-PROT features (organized by feature type) is presented below. An example for each feature is also included to illustrate its use for describing a sequence location or region.

Change Indicators

CONFLICT

Different papers report differing sequences:

```
FT   CONFLICT     304     304          MISSING (IN REF. 3).
```

MUTAGEN

Indicates an experimentally altered site:

```
FT   MUTAGEN       65      65          H->F: 100% ACTIVITY LOSS.
```

VARIANT

Authors report that sequence variants exist:

```
FT   VARIANT      136     136          M -> I.
```

VARSPLIC

Describes sequence variants produced by alternative splicing:

```
FT   VARSPLIC      33      49          MISSING (IN SHORT ISOFORM).
```


Amino Acid Modifications

BINDING

Binding site for chemical group (co-enzyme, prosthetic group, etc.):

```
FT   BINDING       14      14          HEME (COVALENT).
```

CARBOHYD

Glycosylation site:

```
FT   CARBOHYD      53      53          N-LINKED (GLCNAC...) (POTENTIAL).
```

DISULFID

Disulfide bond:

```
FT   DISULFID      23      84          PROBABLE.
```

LIPID

Covalent binding of a lipid moiety:

```
FT   LIPID          2       2          MYRISTATE.
```

Table 3-2 lists the attached groups that are currently defined.

Table 3-2. SWISS-PROT lipid moiety attached groups

Attached group	Description
MYRISTATE	Myristate group attached through an amide bond to the N-terminal glycine residue of the mature form of a protein or to an internal lysine residue.
PALMITATE	Palmitate group attached through a thioether bond to a cysteine residue or through an ester bond to a serine or threonine residue.
FARNESYL	Farnesyl group attached through a thioether bond to a cysteine residue.
GERANYL-GERANYL	Geranyl-geranyl group attached through a thioether bond to a cysteine residue.
GPI-ANCHOR	Glycosyl-phosphatidylinositol (GPI) group linked to the alpha-carboxyl group of the C-terminal residue of the mature form of a protein.
N-ACYL DIGLYCERIDE	N-terminal cysteine of the mature form of a prokaryotic lipoprotein with an amide-linked fatty acid and a glyceryl group to which two fatty acids are linked by ester linkages.

METAL
: Binding site for a metal ion:

```
FT   METAL       28     28       COPPER (POTENTIAL).
```

MOD_RES
: Posttranslational modification of a residue:

```
FT   MOD_RES     686    686      PHOSPHORYLATION (BY PKC).
```

Table 3-3 lists the most frequent modifications.

Table 3-3. Frequently used SWISS-PROT amino acid modifications

Modification	Description
ACETYLATION	N-terminal or other.
AMIDATION	Generally at the C-terminal of a mature active peptide.
BLOCKED	Undetermined N- or C-terminal blocking group.
FORMYLATION	Of the N-terminal methionine.
GAMMA-CARBOXYGLUTAMIC ACID	Of glutamate.
HYDROXYLATION	Of asparagine, aspartic acid, proline or lysine.
METHYLATION	Generally of lysine or arginine.
PHOSPHORYLATION	Of serine, threonine, tyrosine, aspartic acid or histidine.
PYRROLIDONE CARBOXYLIC ACID	N-terminal glutamate which has formed an internal cyclic lactam. This is also called "pyro-Glu".
SULFATION	Generally of tyrosine.

SE_CYS
: Selenocysteine:

```
FT   SE_CYS      52     52
```

THIOETH
: Thioether bond.

THIOLEST
: Thiolester bond.

Regions

CA_BIND
: Extent of a calcium-binding region:

```
FT   CA_BIND     759    770      EF-HAND 1 (POTENTIAL).
```

CHAIN
: Extent of a polypeptide chain in the mature protein:

```
FT   CHAIN       21     119      BETA-2 MICROGLOBULIN.
```

DNA_BIND
: Extent of a DNA-binding region:

```
FT   DNA_BIND    69     128      HOMEOBOX.
```

DOMAIN
: Extent of a domain of interest on the sequence:

```
FT   DOMAIN      22     788      EXTRACELLULAR (POTENTIAL).
```

NP_BIND

Extent of a nucleotide phosphate-binding region:

```
FT   NP_BIND      13      25        ATP.
```

PEPTIDE

Extent of a released active peptide:

```
FT   PEPTIDE      13      107       NEUROPHYSIN 2.
```

PROPEP

Extent of a propeptide:

```
FT   PROPEP       550     574       REMOVED IN MATURE FORM.
```

REPEAT

Extent of an internal sequence repetition:

```
FT   REPEAT       225     307       1.
```

SIGNAL

Extent of a signal sequence (prepeptide).

SIMILAR

Extent of a similarity with another protein sequence:

```
FT   SIMILAR      139     153       STRONG WITH CA-BINDING EF-HAND
SEQUENCE.
```

TRANSIT

Extent of a transit peptide (mitochondrial, chloroplastic, thylakoid, cyanelle or for a microbody):

```
FT   TRANSIT      1       25        MITOCHONDRION.
```

TRANSMEM

Extent of a transmembrane region.

ZN_FING

Extent of a zinc finger region:

```
FT   ZN_FING      319     343       GATA-TYPE.
```

Secondary Structure

Secondary structures are formed as a result of the physical characteristics of the amino acid sidechains of a protein (see Table 3-4).

Table 3-4. SWISS-PROT secondary structure codes

Abbreviation	Description	Type
B	Residue in an isolated beta-bridge	STRAND
E	Hydrogen-bonded beta-strand (extended strand)	STRAND
G	3(10) helix	HELIX
H	Alpha-helix	HELIX
I	Pi-helix	HELIX
S	Bend (five-residue bend centered at residue i)	Not specified
T	H-bonded turn (3-turn, 4-turn or 5-turn)	TURN

For example:

```
FT   HELIX        4       14
```

Others

ACT_SITE

Amino acid(s) involved in the activity of an enzyme:

```
FT   ACT_SITE    193    193       ACCEPTS A PROTON DURING CATALYSIS.
```

INIT_MET

Initiator methionine:

```
FT   INIT_MET      0      0
```

NON_CONS

Non-consecutive residues:

```
FT   NON_CONS   1683   1684
```

NON_TER

The residue at an extremity of the sequence is not the terminal residue:

```
FT   NON_TER     129    129
```

SITE

Any other interesting site on the sequence:

```
FT   SITE        759    760       CLEAVAGE (BY THROMBIN).
```

UNSURE

Uncertainties in the sequence.

References

Bairoch, A., and R. Apweiler. 2000. The SWISS-PROT protein sequence database and its supplement TrEMBL in 2000. *Nucleic Acids Research* 28:45–48.

Main page
: *http://us.expasy.org/sprot/*

Release notes
: *http://us.expasy.org/sprot/relnotes/*

User manual
: *http://us.expasy.org/sprot/userman.html*

Download
: *ftp://us.expasy.org/databases/swiss-prot*

4
Pfam

While many databases are dedicated to organizing protein families and protein domains, Pfam is our preferred database for predicting the function of newly-discovered proteins. Pfam is unique in that it is a manually curated database of protein families derived from protein multiple sequence alignments and profile hidden Markow models. Pfam is a key database for understanding protein function and structure. It is used in many methods, including phylogenetic analysis, secondary structure prediction, and sequence annotation. We're using Pfam Release 7.8.

Pfam Example Flat File

Example 4-1 shows a Pfam flat file. This entry contains terms from the Pfam Field Definitions, discussed later in this chapter.

Example 4-1. Sample Pfam example

```
# STOCKHOLM 1.0
#=GF ID   14-3-3
#=GF AC   PF00244
#=GF DE   14-3-3 proteins
#=GF AU   Finn RD
#=GF AL   Clustalw
#=GF SE   Prosite
#=GF GA   25 25
#=GF TC   35.40 35.40
#=GF NC   19.10 19.10
#=GF BM   hmmbuild -f HMM SEED
#=GF BM   hmmcalibrate --seed 0 HMM
#=GF RN   [1]
#=GF RM   95327195
#=GF RT   Structure of a 14-3-3 protein and implications for
#=GF RT   coordination of multiple signalling pathways.
```

Example 4-1. Sample Pfam example (continued)

```
#=GF RA   Xiao B, Smerdon SJ, Jones DH, Dodson GG, Soneji Y, Aitken
#=GF RA   A, Gamblin SJ;
#=GF RL   Nature 1995;376:188-191.
#=GF RN   [2]
#=GF RM   95327196
#=GF RT   Crystal structure of the zeta isoform of the 14-3-3
#=GF RT   protein.
#=GF RA   Liu D, Bienkowska J, Petosa C, Collier RJ, Fu H, Liddington
#=GF RA   R;
#=GF RL   Nature 1995;376:191-194.
#=GF RN   [3]
#=GF RM   96182649
#=GF RT   Interaction of 14-3-3 with signaling proteins is mediated
#=GF RT   by the recognition of phosphoserine.
#=GF RA   Muslin AJ, Tanner JW, Allen PM, Shaw AS;
#=GF RL   Cell 1996;84:889-897.
#=GF RN   [4]
#=GF RM   97424374
#=GF RT   The 14-3-3 protein binds its target proteins with a common
#=GF RT   site located towards the C-terminus.
#=GF RA   Ichimura T, Ito M, Itagaki C, Takahashi M, Horigome T,
#=GF RA   Omata S, Ohno S, Isobe T
#=GF RL   FEBS Lett 1997;413:273-276.
#=GF RN   [5]
#=GF RM   96394689
#=GF RT   Molecular evolution of the 14-3-3 protein family.
#=GF RA   Wang W, Shakes DC
#=GF RL   J Mol Evol 1996;43:384-398.
#=GF RN   [6]
#=GF RM   96300316
#=GF RT   Function of 14-3-3 proteins.
#=GF RA   Jin DY, Lyu MS, Kozak CA, Jeang KT
#=GF RL   Nature 1996;382:308-308.
#=GF DR   PROSITE; PDOC00633;
#=GF DR   SMART; 14_3_3;
#=GF DR   PRINTS; PR00305;
#=GF DR   SCOP; 1a4o; fa;
#=GF DR   PDB; 1a37 A; 3; 228;
#=GF DR   PDB; 1a37 B; 3; 228;
#=GF DR   PDB; 1a38 A; 3; 228;
#=GF DR   PDB; 1a38 B; 3; 228;
#=GF DR   PDB; 1a4o A; 3; 228;
#=GF DR   PDB; 1a4o B; 3; 228;
#=GF DR   PDB; 1a4o C; 3; 228;
#=GF DR   PDB; 1a4o D; 3; 228;
#=GF DR   PDB; 1qja B; 3; 229;
#=GF DR   PDB; 1qja A; 3; 230;
#=GF DR   PDB; 1qjb A; 3; 232;
#=GF DR   PDB; 1qjb B; 3; 232;
#=GF DR   INTERPRO; IPR000308;
#=GF SQ   148
#=GS O61131/11-251        AC O61131
```

Example 4-1. Sample Pfam example (continued)

```
<deleted for brevity>

#=GS 143Z_HUMAN/3-236 DR PDB; 1qjb B; 3; 232;
O61131/11-251                RSDCTYRSKLAEQAERYDEMADAMRTLVEQCVnn.......
dkdELTVEERNLLSVAYKNAVGARRASWRIISSVEQKEMSKA.NVHNKNIAATYRKKVEEELNNIC.QDILN.
LLTKKLIPNT..SESESKVFYYKMKGDYYRYISEFS.CDE.
GKKEASNFAQEAYQKATDIAENELPSTHPIRLGLALNYSVFFY..EILNQPHQACEMAKRAF...DDAITEFDNV..
SEDS..YKDSTLI.MQLLRDNLTLWTSDLQGDQ

<deleted for brevity>

Q9XZV0/2-235                 KEELLNRCKLNDLIENYGEMFEYLKELSHIKI...........
DLQPDELDLITRCTKCYIGHKRGQYRKILTLIDKDKIVD.NQKNSALLEILRKKLSEEILLLC.NSTIE.LSQNFLNNNV.
.FPKKTQLFFTKIIADHYRYIYEIN.GKE.DIKLKAKEYYE--KGLQTIKTCKYNSTETAYLTFYLNYSVFLH..
DTMRNTEESIKVSKACL...YEALKDTEDI..VDNS..QKDIVLL.CQMLKDNISLWKTETNEDN
#=GC SS_cons                 HHHHHHHHHHHHHTTCHHHHHHHHHHHHHTTSC...........
CCCHHHHHHHHHHHHHHHHHHHHHHHHHHHHHHCTTT--.CCHHHHHHHHHHHHHHHHHHHHHH.HHHHH.HHHHTTTTCC.
.CSCHHHHHHHHHHHHHHHHHHHHHC.CSC.HHHHHHHHHHHHHHHHHHHHHHCHCCTTCHCHHHHHHHHHHHHC..
HTSCCHHHCHHHHHHHH...HHHHTTCGGC..CTTT..HHHHHHH.HHHHHHHHHHCTCCCXXXX
#=GC SA_cons                 2631032030035051251005002200335 2...........
40455004001200330023104024201521791 79--.38752510440144014203510.43002.0035201642.
.754403000010100011100201.867.74651253025003402520676351131221000 01001127..
31372485135106412...5415867932..3994..6651462.142043126627759XXXX
//
```

Pfam Field Definitions

The field codes found in a Pfam flat file help display information for human readabilty and machine-based parsing. A typical entry contains several two-letter Pfam field codes. Table 4-1 provides definitions and descriptions of these codes.

Table 4-1. Pfam field definitions

Field	Definition	Description
AC	Accession number	PFxxxxx or PBxxxxxx.
ID	Identification	15 characters or less.
DE	Definition	80 characters or less.
AU	Author	Author of the entry.
AL	Alignment method of seed	Method used to align the seed members. Approved AL lines are: Clustalv Clustalw Clustalw_mask_xxxx Domainer HMM_built_from_alignment HMM_simulated_annealing Manual Prosite_pattern Prodom Structure_superposition pftools Unknown

Table 4-1. Pfam field definitions (continued)

Field	Definition	Description
BM	HMM building command lines	
SE	Source of seed	The source suggesting seed members belong to a family.
GA	Gathering threshold	Search threshold to build the full alignment.
NC	Noise cutoff	This field refers to the bit scores of the highest scoring match not in the full alignment.
TC	Trusted cutoff	This field refers to the bit scores of the lowest scoring match in the full alignment.
TP	Type field	The type field is a compulsory field describing the type of family. At present it can be one of: Family Domain Repeat Motif
PI	Previous IDs	
DC	Database Comment	Comment for database reference.
DR	Database Reference	Reference to external database.
RC	Reference Comment	Comment for literature reference.
RN	Reference Number	Digit in square brackets.
RM	Reference Medline	Eight digit number.
RT	Reference Title	Title of paper.
RA	Reference Author	Author of paper.
RL	Reference Location	Location of paper.
CC	Comment	Comment lines provide annotation and other information.
NE	Pfam accession	Indicated those cases where there is a nested domain.
SQ	Sequence	Nr of sequences, start of alignment.
//	End of alignment	

References

Bateman, A., E. Birney, L. Cerruti, R. Durbin, L. Etwiller, S. R. Eddy, S. Griffiths-Jones, K. L. Howe, M. Marshall, and E. L. L. Sonnhammer. 2002. The Pfam Protein Families Database. *Nucleic Acids Research* 30 (1):275–280.

Main page
: *http://pfam.wustl.edu/*

Release notes
: *ftp://ftp.genetics.wustl.edu/pub/Pfam/relnotes.txt*

Help pages
: *http://pfam.wustl.edu/help.shtml*

Download
: *ftp://ftp.genetics.wustl.edu/pub/Pfam/*

5
PROSITE

PROSITE is one the many popular databases for sequence profiles, patterns, and motifs. It is one of our favorite databases for sequence analysis, and we hope you find it as useful. The database was created so that computer-based tools could quickly identify sequences containing any known protein motif. PROSITE patterns represent another key database for basic sequence analysis and protein function determination. We're using PROSITE Release 17.

PROSITE Example Flat File

Example 5-1 contains a sample pattern entry from a PROSITE flat file. This entry contains examples of the PROSITE Field Definitions, discussed later in this chapter.

Example 5-1. Sample PROSITE pattern entry

```
ID   PPASE; PATTERN.
AC   PS00387;
DT   NOV-1990 (CREATED); NOV-1995 (DATA UPDATE); NOV-1995 (INFO UPDATE).
DE   Inorganic pyrophosphatase signature.
PA   D-[SGN]-D-[PE]-[LIVM]-D-[LIVMGC].
NR   /RELEASE=32,49340;
NR   /TOTAL=16(16); /POSITIVE=11(11); /UNKNOWN=0(0); /FALSE_POS=5(5);
NR   /FALSE_NEG=0; /PARTIAL=2;
CC   /TAXO-RANGE=A?EP?; /MAX-REPEAT=1;
CC   /SITE=1,magnesium; /SITE=3,magnesium; /SITE=6,magnesium;
DR   P21216, IPYR_ARATH, T; P37980, IPYR_BOVIN, T; P17288, IPYR_ECOLI, T;
DR   P44529, IPYR_HAEIN, T; P13998, IPYR_KLULA, T; P19117, IPYR_SCHPO, T;
DR   P37981, IPYR_THEAC, T; P19514, IPYR_THEP3, T; P38576, IPYR_THETH, T;
DR   P00817, IPYR_YEAST, T; P28239, IPY2_YEAST, T;
DR   P19371, IPYR_DESVH, P; P21616, IPYR_PHAAU, P;
DR   P09167, AERA_AERHY, F; P12351, CYP1_YEAST, F; P24653, Y101_NPVOP, F;
DR   P37904, YCEI_ECOLI, F; P39303, YJFU_ECOLI, F;
```

Example 5-1. Sample PROSITE pattern entry (continued)

```
3D   1PYP;
DO   PDOC00325;
//
```

Example 5-2 contains a sample profile (matrix) entry from a PROSITE flat file. This entry contains further examples of the PROSITE Field Definitions described later in this chapter.

Example 5-2. Sample PROSITE profile (matrix)

```
ID   GLOBIN; MATRIX.
AC   PS01033;
DT   JUN-1994 (CREATED); DEC-2001 (DATA UPDATE); DEC-2001 (INFO UPDATE).
DE   Globins profile.
MA   /GENERAL_SPEC: ALPHABET='ABCDEFGHIKLMNPQRSTVWYZ'; LENGTH=154;
MA   /DISJOINT: DEFINITION=PROTECT; N1=1; N2=154;
MA   /NORMALIZATION: MODE=1; FUNCTION=LINEAR; R1=-0.8705306; R2=0.0209303;
TEXT='-LogE';
MA   /CUT_OFF: LEVEL=0; SCORE=424; N_SCORE=8.0; MODE=1; TEXT='!';
MA   /CUT_OFF: LEVEL=-1; SCORE=353; N_SCORE=6.5; MODE=1; TEXT='?';
MA   /DEFAULT: D=-20; I=-20; MI=-210; MD=-210; IM=0; DM=0;
MA   /I: I=-6;
MA   /M: SY='A'; M=7,-7,-8,-10,-10,-8,3,-12,-4,-8,-6,-4,-6,-10,-10,-10,3,4,3,-
14,-10,-10; D=-6;
MA   /I: I=-6; MI=-59; MD=-59;
MA   /M: SY='H'; M=1,-3,-21,0,-6,-20,0,2,-16,-10,-16,-10,-4,0,-8,-12,-2,-9,-11,-
23,-13,-8; D=-6;

<deleted for brevity>

MA   /M: SY='H'; M=-1,4,-18,5,3,-19,-10,9,-20,8,-16,-9,3,-10,2,8,-2,-7,-14,-19,-
6,1; D=-5;
MA   /I: I=0; MI=*;
NR   /RELEASE=40.7,103373;
NR   /TOTAL=797(796); /POSITIVE=796(795); /UNKNOWN=0(0); /FALSE_POS=1(1);
NR   /FALSE_NEG=0; /PARTIAL=3;
CC   /MATRIX_TYPE=protein_domain;
CC   /SCALING_DB=reversed;
CC   /AUTHOR=P_Bucher;
CC   /TAXO-RANGE=??EP?; /MAX-REPEAT=9;
CC   /FT_KEY=DOMAIN; /FT_DESC=GLOBIN;
DR   P04252, BAHG_VITST, T; Q03331, FHP_CANNO , T; P39676, FHP_YEAST , T;
DR   P02212, GLB1_ANABR, T; P19363, GLB1_ARTSX, T; P14805, GLB1_CALSO, T;
DR   P02221, GLB1_CHITH, T; P02216, GLB1_GLYDI, T; P20412, GLB1_LAMSP, T;
DR   P41260, GLB1_LUCPE, T; P08924, GLB1_LUMTE, T; P21197, GLB1_MORMR, T;

<deleted for brevity>

DR   P42430, YKYB_BACSU, F;
3D   1VHB; 2VHB; 3VHB; 1HBG; 2HBG; 1B0B; 1EBT; 1FLP; 1MOH; 1HBI; 2HBI; 3HBI;
3D   3SDH; 4HBI; 4SDH; 5HBI; 6HBI; 7HBI; 1ECA; 1ECD; 1ECN; 1ECO; 1VRE; 1VRF;
3D   2LHB; 3LHB; 1DM1; 1MBA; 2FAL; 2FAM; 3MBA; 4MBA; 5MBA; 1SCT; 1HLB; 1HLM;
```

Example 5-2. Sample PROSITE profile (matrix) (continued)

```
3D   1OUT; 1OUU; 1A4F; 1FSX; 1HDA; 1CG5; 1CG8; 1IBE; 2DHB; 2MHB; 1A00; 1A01;
3D   1A0U; 1A0V; 1A0W; 1A0X; 1A0Y; 1A0Z; 1A3N; 1A3O; 1A9W; 1ABW; 1ABY; 1AJ9;
3D   1AXF; 1B86; 1BAB; 1BBB; 1BIJ; 1BUW; 1BZ0; 1BZ1; 1BZZ; 1CLS; 1CMY; 1COH;
3D   1DSH; 1DXT; 1DXU; 1DXV; 1FDH; 1GBU; 1GBV; 1GLI; 1HAB; 1HAC; 1HBA; 1HBB;
3D   1HBS; 1HCO; 1HDB; 1HGA; 1HGB; 1HGC; 1HHO; 1NIH; 1QI8; 1QSH; 1QSI; 1RVW;
3D   1SDK; 1SDL; 1THB; 1VWT; 2HBC; 2HBD; 2HBE; 2HBF; 2HBS; 2HCO; 2HHB; 2HHD;
3D   2HHE; 3HHB; 4HHB; 6HBW; 1SPG; 1HDS; 1HBH; 1PBX; 1QPW; 2PGH; 1HBR; 1CBL;
3D   1CBM; 1ITH; 1D8U; 1CQX; 1GDI; 1GDJ; 1GDK; 1GDL; 1LH1; 1LH2; 1LH3; 1LH5;
3D   1LH6; 1LH7; 2GDM; 2LH1; 2LH2; 2LH3; 2LH5; 2LH6; 2LH7; 1BIN; 1FSL; 1LHS;
3D   1LHT; 1EMY; 1MBS; 1AZI; 1BJE; 1DWR; 1DWS; 1DWT; 1HRM; 1HSY; 1RSE; 1WLA;
3D   1XCH; 1YMA; 1YMB; 1YMC; 2MM1; 101M; 102M; 103M; 104M; 105M; 106M; 107M;
3D   108M; 109M; 110M; 111M; 112M; 1A6G; 1A6K; 1A6M; 1A6N; 1ABS; 1AJG; 1AJH;
3D   1BVC; 1BVD; 1BZ6; 1BZP; 1BZR; 1CH1; 1CH2; 1CH3; 1CH5; 1CH7; 1CH9; 1CIK;
3D   1CIO; 1CO8; 1CO9; 1CPO; 1CP5; 1CPW; 1CQ2; 1DO1; 1DO3; 1DO4; 1DO7; 1DTI;
3D   1DTM; 1DUK; 1DUO; 1DXC; 1DXD; 1EBC; 1F63; 1F65; 1F6H; 1FCS; 1HJT; 1IOP;
3D   1IRC; 1JDO; 1LTW; 1MBC; 1MBD; 1MBI; 1MBN; 1MBO; 1MCY; 1MGN; 1MLF; 1MLG;
3D   1MLH; 1MLJ; 1MLK; 1MLL; 1MLM; 1MLN; 1MLO; 1MLQ; 1MLR; 1MLS; 1MLU; 1MOA;
3D   1MOB; 1MOC; 1MOD; 1MTI; 1MTJ; 1MTK; 1MYF; 1MYM; 1OBM; 1OFJ; 1OFK; 1SPE;
3D   1SWM; 1TES; 1VXA; 1VXB; 1VXC; 1VXD; 1VXE; 1VXF; 1VXG; 1VXH; 1YOG; 1YOH;
3D   1YOI; 2CMM; 2MB5; 2MBW; 2MGA; 2MGB; 2MGC; 2MGD; 2MGE; 2MGF; 2MGG; 2MGH;
3D   2MGI; 2MGJ; 2MGK; 2MGL; 2MGM; 2MYA; 2MYB; 2MYC; 2MYD; 2MYE; 2SPL; 2SPM;
3D   2SPN; 2SPO; 4MBN; 5MBN; 1M6C; 1M6M; 1MDN; 1MNH; 1MNI; 1MNJ; 1MNK; 1MNO;
3D   1MWC; 1MWD; 1MYG; 1MYH; 1MYI; 1MYJ; 1PMB; 1YCA; 1YCB; 1MYT; 1ASH;
DO   PDOC00793;
//
```

PROSITE

PROSITE Field Definitions

The field codes found in a PROSITE flat file help to arrange the information for human readability and machine-based parsing. There are several PROSITE field codes found in an entry; each is represented with a two-letter abbreviation. Table 5-1 provides definitions and descriptions of these field codes.

Table 5-1. PROSITE field definitions

Field	Definition	Description
ID	Identification	The second item indicates the type of entry: PATTERN MATRIX RULE
AC	Accession number	PSnnnnn.
DT	Date	Date of entry or last modification of the entry.
DE	Short description	Descriptive information about the entry content.
PA	Pattern	The definition of a PROSITE pattern.
MA	Matrix/profile	The definition of a PROSITE profile/matrix.
RU	Rule	The definition of a PROSITE rule.

Table 5-1. PROSITE field definitions (continued)

Field	Definition	Description
NR	Numerical results	This contain information relevant to the results of the scan with a pattern on the complete SWISS-PROT knowledgebase. The following qualifiers are used: /RELEASE /TOTAL /POSITIVE /UNKNOWN /FALSE_POS /FALSE_NEG /PARTIAL
CC	Comments	Various types of comments. The following qualifiers are used. /TAXO-RANGE /MAX-REPEAT /SITE /SKIP-FLAG /MATRIX_TYPE /SCALING_DB /AUTHOR /FT_KEY /FT_DESC
DR	Cross-references to SWISS-PROT	These are used as pointers to SWISS-PROT entries.
3D	Cross-references to PDB	These are used to list the Protein Data Bank entries.
DO	Pointer to the documentation file	This contains a pointer to the entry in the PROSITE documentation file that describes the entry.
//	Termination line	This designates the end of an entry.

References

Falquet L., Pagni M., Bucher P., Hulo N., Sigrist C.J.A., Hofmann K., Bairoch A. 2002. The PROSITE database, its status in 2002. *Nucleic Acids Research*. Jan 1;30(1):235–8.

Sigrist CJA, Cerutti L, Hulo N, Gattiker A, Falquet L, Pagni M, Bairoch A, Bucher P. 2002. PROSITE: a documented database using patterns and profiles as motif descriptors. *Brief Bioinform*. Sep;3(3):265–74.

Main page
: *http://us.expasy.org/prosite/*

Release notes
: *http://us.expasy.org/prosite/psrelnot.html*

User manual
: *http://us.expasy.org/prosite/prosuser.html*

Download
: *ftp://us.expasy.org/databases/prosite/*

Tools

Now that we've described the common data formats and databases, it's time to get to work! What can you do with the data? You can compare two or more sequences, compute properties for the sequences, and look for patterns and subsequences. The possibilities are nearly limitless.

Since there's no way to describe all of the available tools—or even just the ones we use—we decided to showcase the tools we use most. We've included the classics: Readseq, BLAST, ClustalW, and HMMER. We also cover MEME and MAST (two tools that deserve to be better known), and a rising star called BLAT. The final chapter contains a wealth of information about the widely used open source suite of EMBOSS tools.

Each tool's brief description is followed by examples and command-line options. We've also included helpful web sites and other references.

We'd like to thank all the developers for making this rich abundance of documentation available to users!

6
Readseq

Readseq is a classic, dating from 1989. Developed by Don Gilbert, this program reads and writes nucleotide and protein sequences in many useful formats. The Java version is the most current; we're using Version 2.

To run Readseq use:

```
java -cp readseq.jar run options inputfiles
```

For more details use:

```
java -cp readseq.jar help more
```

This chapter contains a list of the command line options used in Readseq.

Supported Formats

Table 6-1 contains the formats supported by Readseq. **ID** is a number that can be used for this format (name is prefered). Alternate **Names** are separated by using the | character. You can use any of these names to specify a format. **R** and **W** indicate if Readseq can read and write this format. **I** means the format is interleaved. **F** indicates that sequence record documentation and features are parsed. **S** indicates that the format contains sequence data. **Content-type** is the magic string sent for that format through a CGI web server. The **suffix** is the standard file suffix used for that format.

Table 6-1. Supported formats for Readseq

ID	Name	R	W	I	F	S	Content-type	Suffix
1	GenBank|gb	T	T	F	T	T	biosequence/genbank	*.gb*
2	EMBL|em	T	T	F	T	T	biosequence/embl	*.embl*
3	Pearson|Fasta|fa	T	T	F	F	T	biosequence/fasta	*.fasta*
4	GCG	T	T	F	F	T	biosequence/gcg	*.gcg*

Table 6-1. Supported formats for Readseq (continued)

ID	Name	R	W	I	F	S	Content-type	Suffix
5	MSF	T	T	T	F	T	biosequence/msf	*.msf*
6	Clustal	T	T	T	F	T	biosequence/clustal	*.aln*
7	NBRF	T	T	F	F	T	biosequence/nbrf	*.nbrf*
8	PIR|CODATA	T	T	F	F	T	biosequence/codata	*.pir*
9	ACEDB	T	T	F	F	T	biosequence/acedb	*.ace*
10	Phylip3.2	T	T	T	F	T	biosequence/phylip2	*.phylip2*
11	Phylip|Phylip4	T	T	T	F	T	biosequence/phylip	*.phylip*
12	Plain|Raw	T	T	F	F	T	biosequence/plain	*.seq*
13	PAUP|NEXUS	T	T	T	F	T	biosequence/nexus	*.nexus*
14	XML	T	T	F	T	T	biosequence/xml	*.xml*
15	FlatFeat|FFF	T	T	F	T	F	biosequence/fff	*.fff*
16	GFF	T	T	F	T	F	biosequence/gff	*.gff*
17	BLAST	T	F	T	F	T	biosequence/blast	*.blast*
18	Pretty	F	T	T	F	T	biosequence/pretty	*.pretty*
19	SCF	T	F	F	F	T	biosequence/scf	*.scf*
20	DNAStrider	T	T	F	F	T	biosequence/strider	*.strider*
21	IG|Stanford	T	T	F	F	T	biosequence/ig	*.ig*
22	Fitch	F	F	F	F	T	biosequence/fitch	*.fitch*
23	ASN.1	F	F	F	F	T	biosequence/asn1	*.asn*

Command-Line Options

Tables 6-2 through 6-6 summarize Readseq's command-line options.

Table 6-2. Primary pptions

Option	Definition
-a[ll]	Select all sequences. "all" causes processing of all sequences (default now for Version 2, for compatibility with version 1). Use" items=1,2,3" to select a subset.
-c[aselower]	Change to lower case. "caselower" and "CASEUPPER" will convert sequence case.
-C[ASEUPPER]	Change to UPPERCASE.
-degap[=-]	Remove gap symbols. "degap=symbol" will remove this symbol from output sequence (- normally).
-f[ormat=]#	Format number for output.
-f[ormat=]Name	Format name for output. See formats list (Table 6-1) for names and numbers. "format=genbank", "format=gb", "format=xml", etc., selects an output format. You can also use format number, but these numbers may change with revisions. Alternate names of formats are listed in Table 6-1. "Pearson|FastA|fa" allows "pearson", "fasta", or "fa" as a name). This is case-insensitive.
-inform[at]=#	Input format number.
-inform[at]=Name	Input format name. Assume input data is this format. "inform=genbank" lets you specify data input format. Normally Readseq guesses the input format (usually correctly). Use this option if you wish to bypass this input format guessing.

Table 6-2. Primary pptions (continued)

Option	Definition
-i[tem=2,3,4]	Select Item number(s) from several. "items=2,3,4" will select these sequence records from a multisequence input file.
-l[ist]	List sequences only. "list" will list titles of sequence records.
-o[utput=]out.seq	Redirect Output. "output=file", sends output to named file.
-p[ipe]	Pipe (command line, < stdin, > stdout). "pipe" will cause input data to come from STDIN and output go to STDOUT Unix standard files (unless -out is given and input file given), and no prompting or progress reports will occurr.
-r[everse]	Reverse-complement of input sequence. "reverse" will write the sequence from end to start, and DNA bases are complemented. Amino residues are not complemented.
-t[ranslate=]io	Translate input symbol [i] to output symbol [o]. Use several -tio to translate several symbols translates given sequence bases, e.g., -tAN to change "A" to "N".
-v[erbose]	Verbose progress. "verbose" will print some progress reports.
-ch[ecksum]	Calculate & print checksum of sequences.

Table 6-3. Documentation and feature table extraction options

Option	Definition
-feat[ures]=exon,CDS...	Extract sequence of selected features.
-nofeat[ures]=repeat_region,intron...	Remove sequence of selected features. "feature=CDS,intron" lets you specify those features to extract, or remove, in the output. Currently this causes each feature to produce a new sequence record.
-field=AC,ID...	Include selected document fields in output.
-nofield=COMMENT,...	Remove selected document fields from output.

Table 6-4. Subrange options

Option	Definition
-subrange=-1000..10	Extract subrange of sequence for feature locations: -subrange=1..end -subrange=end-10..end+99
-extract=10000..99999	Extract all features and sequence from given base range.

Table 6-5. Pair, unpair options

Option	Definition
-pair=1	Combine features (fff,gff) and sequence files to one output.
-unpair=1	Split features, sequence from one input to two files.

Table 6-6. Pretty format options

Option	Definition
-wid[th]=#	Sequence line width.
-tab=#	Left indent.
-col[space]=#	Column space within sequence line on output.

Table 6-6. Pretty format options (continued)

Option	Definition
-gap[count]	Count gap chars in sequence numbers.
-nameleft, -nameright[=#]	Name on left/right side [=max width].
-nametop	Name at top/bottom.
-numleft, -numright	Seq index on left/right side.
-numtop, -numbot	Index on top/bottom.
-match[=.]	Use match base for 2..n species.
-inter[line=#]	Blank line(s) between sequence blocks.

References

Gilbert, D. G. 1999. Readseq Version 2, an improved biosequence conversion tool, written in the Java language. *Bionet.Software* (August).

Main page

http://iubio.bio.indiana.edu/soft/molbio/readseq/

README

http://iubio.bio.indiana.edu/soft/molbio/readseq/Readme

Download

http://iubio.bio.indiana.edu/soft/molbio/readseq/classic/

http://iubio.bio.indiana.edu/soft/molbio/readseq/java/

BLAST

BLAST (Basic Local Alignment Search Tool) is probably the best-known program in sequence analysis. It compares two sequences by trying to align them, and is also used to lookup sequences in a database. The algorithm starts by looking for exact matches, then expands the aligned regions by allowing for mismatches. For details, see the "References" section at the end of this chapter.

This chapter contains a guide to the command-line options used in BLAST programs. The programs are listed in the order you might expect to use them. Each entry includes a brief program description, a command-line entry example, and a table summarizing any available options. We're using Version 2.2.5 of BLAST.

formatdb

formatdb is used to format protein or nucleotide source databases before these databases can be searched by *blastall*, *blastpgp* or *MegaBLAST*.

An example of a *formatdb* command-line entry:

```
formatdb -i fastafile -p F -oflat file T
```

The following table summarizes the *formatdb* options.

Option	Definition	Type	Default
-a	Input file is database in ASN.1 format (otherwise FASTA is expected).	[T/F]	F
-b	ASN.1 database in binary mode: T = Binary. F = Text mode.	[T/F]	F
-e	Input is a Seq-entry.	[T/F]	F
-i	Input file for formatting (this parameter must be set).	[File In]	
-l	Logfile name.	[File Out]	formatdb.log
-n	Base name for BLAST files.	[String]	

Option	Definition	Type	Default
-o	Parse options: T = True: Parse SeqId and create indexes. F = False: Do not parse SeqId. Do not create indexes.	[T/F]	F
-p	Type of file: T = Protein. F = Nucleotide.	[T/F]	T
-s	Create indexes limited only to accessions - sparse.	[T/F]	F
-t	Title for database file.	[String]	
-v	Number of sequence bases to be created in the volume.	[Integer]	0
-A	Create ASN.1 structured deflines.	[T/F]	F
-B	Binary GIfile produced from the GIfile. This option should be used with the -F option.	[File Out]	
-F	GIfile (file containing list of GIs).	[File In]	
-L	Create an alias file with this name.	[File Out]	

blastall

blastall allows use of all BLAST programs (*blastn*, *blastp*, *blastx*, *tblastx*, and *tblastn*). The following table summarizes the query, database sequence, and alignment types for the various BLAST commands.

Program	Query sequence type	Database sequence type	Alignment sequence type
blastn	nucleotide	nucleotide	nucleotide
blastp	protein	protein	protein
blastx	nucleotide	protein	protein
tblastn	protein	nucleotide	protein
tblastx	nucleotide	nucleotide	protein

An example of a *blastall* command-line entry:

```
blastall -p programname -d databasefilename -i queryfilename -o
outputfilename
```

The following table summarizes the *blastall* options.

Option	Definition	Type	Default
-a	Number of processors to use.	[Integer]	1
-b	Number of database sequence to show alignments for (B).	[Integer]	250
-d	Database: multiple database names are bracketed by quotations, for example : -d "db1 db2 db3".	[String]	nr
-e	Expectation value (E).	[Real]	10.0
-f	Threshold for extending hits, default if 0.	[Integer]	0
-g	Perfom gapped alignment (not available with *tblastx*).	[T/F]	T
-i	Query File.	[File In]	stdin
-l	Restrict search of database to list of GIs.	[String]	Optional

Option	Definition	Type	Default
-m	Alignment view options: 0 = Tairwise. 1 = Query-anchored, showing identities. 2 = Query-anchored, no identities. 3 = Flat query-anchored, show identities. 4 = Flat query-anchored, no identities. 5 = Query-anchored, no identities and blunt ends. 6 = Flat query-anchored, no identities and blunt ends. 7 = XML Blast output. 8 = Tabular.	[Integer]	0
-n	MegaBLAST search.	[T/F]	F
-o	BLAST report Output File.	[File Out]	stdout
-p	Program Name.	[String]	
-q	Penalty for a nucleotide mismatch (*blastn* only).	[Integer]	-3
-r	Reward for a nucleotide match (*blastn* only).	[Integer]	1
-v	Number of database sequences to show one-line descriptions for (V).	[Integer]	500
-y	Dropoff (X) for blast extensions in bits (0.0 invokes default behavior).	[Real]	0.0
-z	Effective length of the database (use 0 for the real size).	[Real]	0
-A	Multiple Hits window size (0 for single-hit algorithm).	[Integer]	40
-D	DB Genetic code (for *tblast*[*nx*] only).	[Integer]	1
-E	Cost to extend a gap (0 invokes default behavior).	[Integer]	0
-F	Filter query sequence (DUST with *blastn*, SEG with others).	[String]	T
-G	Cost to open a gap (0 invokes default behavior).	[Integer]	0
-I	Show GIs in deflines.	[T/F]	F
-J	Believe the query defline.	[T/F]	F
-K	Number of best hits from a region to keep (off by default; if used, a value of 100 is recommended).	[Integer]	0
-L	Location on query sequence.	[String]	
-M	Matrix.	[String]	BLOSUM62
-O	SeqAlign file.	[File Out]	
-P	0 = Multiple hits, 1-pass. 1 = Single hit, 1-pass. 2 = 2-pass.	[Integer]	0
-Q	Query Genetic code to use.	[Integer]	1
-R	PSI-TBLASTN checkpoint file.	[File In]	
-S	Query strands to search against database (for *blast*[*nx*], and *tblastx*). 3 is both, 1 is top, 2 is bottom.	[Integer]	3
-T	Produce HTML output.	[T/F]	F
-U	Use lower case filtering of FASTA sequence.	[T/F]	F
-W	Word size, default if 0.	[Integer]	0
-X	X dropoff value for gapped alignment (in bits) (0 invokes default behavior).	[Integer]	0
-Y	Effective length of the search space (use 0 for the real size).	[Real]	0
-Z	X dropoff value for final gapped alignment (in bits).	[Integer]	0

megablast

megablast uses an algorithm for nucleotide sequence alignment searches and concatenates many queries to decrease the amount of time spent scanning the database.

An example of a *megablast* command-line entry:

```
megablast -d databasefilename -i queryfilename -D 2 -o outputfilename
```

The following table summarizes the MegaBLAST options.

Option	Definition
-b	Maximal number of reported alignments for a given database sequence. This option is meaningful only in conjunction with -D 2.
-e	The cutoff expectation value.
-f	Show full IDs in the output.
-p	Cutoff by percentage of identity.
-s	Minimal hit score to report. By default this value is set to W.
-v	Maximal number of database sequences to report alignments from. This option is meaningful only in conjunction with -D 2.
-D	Type of the MegaBLAST output: 0 Produce one-line output for each alignment, in the form subject-ID==<[+-]query-ID> (s_off q_off s_ end q_end) score. 1 Show the same output as level 0, plus the endpoints and percentage of identical nucleotides for each ungapped segment in the alignment. 2 Show the traditional BLAST (*blastn*) output. 3 Show one-line output for each alignment, with the following fields tab-separated: Query ID, Subject ID, percent of identity, alignment length, number of mismatches (not including gaps), number of gap openings, start of alignment in query, end of alignment in query, start of alignment in subject, end of alignment in subject, expected value, bit score.
-F	Filtering. The available filters for nucleotide BLAST or MegaBLAST searches are: D = Dust. R = Human repeats. V = Vector screen. L = Low complexity (equivalent to D). Finally, if letter "m" is included in the filter string, all types of filters are used to mask the query sequence regions only on the word finding stage and do not affect the extension stage.
-G, -E	Affine gapping penalties. The affine version of MegaBLAST requires significantly more memory, so it should be avoided if possible, especially when some of the query or database sequences are very long.
-J	Believe the query defline. The default is T (TRUE) for all types of output except -D 2. Note: If the sequence IDs in the FASTA file are not unique, this option must be set to F (FALSE).
-M	Maximal total length of queries to be concatenated for a single MegaBLAST search.
-O	ASN.1 *seqalign* file. It is only meaningful in conjunction with -D 2.
-P	Maximal number of positions for a hash value. This can be useful when running very long unmasked sequences.
-Q	Masked query output. The output is written to a file specified by the -Q option. It can be used only in onjunction with -D 2.
-U	Use lower case filtering of FASTA sequences. The default for this option is set to FALSE.
-W	Word size.
-X	X-dropoff value.

blastpgp

blastpgp performs gapped *blastp* searches and can be used to perform iterative searches in psi-blast and phi-blast mode.

An example of a *blastpgp* command-line entry:

```
blastpgp -i queryfilename -B alignmentfilename -j 2 -d databasefilename
```

The following table summarizes the *blastpgp* options.

Option	Definition	Type	Default
-a	Number of processors to use.	[Integer]	1
-b	Number of database sequence to show alignments for (B).	[Integer]	250
-c	Constant in pseudocounts for multipass version.	[Integer]	9
-d	Database.	[String]	nr
-e	Expectation value (E).	[Real]	10.0
-f	Threshold for extending hits.	[Integer]	0
-g	Gapped.	[T/F]	T
-h	e-value threshold for inclusion in multipass model.	[Real]	0.005
-i	Query File.	[File In]	stdin
-j	Maximum number of passes to use in multipass version.	[Integer]	1
-k	Hit File for PHI-BLAST.	[File In]	hit_file
-l	Restrict search of database to list of GIs.	[String]	
-m	Alignment view options: 0 = Pairwise 1 = Query-anchored, showing identities. 2 = Query-anchored, no identities. 3 = Flat query-anchored, show identities. 4 = Flat query-anchored, no identities. 5 = qQuery-anchored no identities and blunt ends. 6 = Flat query-anchored, no identities and blunt ends. 7 = XML Blast output. 8 = Tabular output.	[Integer]	0
-o	Output file for alignment.	[File Out]	stdout
-p	Program option for PHI-BLAST.	[String]	blastpgp
-s	Compute locally optimal Smith-Waterman alignments.	[T/F]	F
-t	Tweak Lambda, K, and score matrix for each match.	[T/F]	T
-v	Number of database sequences to show one-line descriptions for (V).	[Integer]	500
-y	Dropoff (X) for blast extensions in bits (default if 0).	[Real]	7.0
-z	Effective length of the database (use 0 for the real size).	[Integer]	0
-A	Multiple hits window size (0 for single-hit algorithm).	[Integer]	40
-B	Input alignment file for PSI-BLAST restart.	[File In]	
-C	Output file for PSI-BLAST checkpointing.	[File Out]	
-E	Cost to extend a gap.	[Integer]	1
-F	Filter query sequence with SEG.	[String]	F
-G	Cost to open a gap.	[Integer]	11

Option	Definition	Type	Default
-H	End of required region in query (-1 indicates end of query).	[Integer]	-1
-I	Show GIs in deflines.	[T/F]	F
-J	Believe the query defline.	[T/F]	F
-K	Number of best hits from a region to keep.	[Integer]	0
-L	Cost to decline alignment (disabled when 0).	[Integer]	0
-M	Matrix.	[String]	BLOSUM62
-N	Number of bits to trigger gapping.	[Real]	22.0
-O	SeqAlign file ("Believe the query defline" must be TRUE).	[File Out]	Optional
-P	0 = Multiple hits, 1-pass. 1 = Single hit, 1-pass. 2 = 2-pass.	[Integer]	0
-Q	Output file for PSI-BLAST matrix in ASCII.	[File Out]	
-R	Input file for PSI-BLAST restart.	[File In]	
-S	Start of required region in query.	[Integer]	1
-T	Produce HTML output.	[T/F]	F
-U	Use lowercase filtering of FASTA sequence.	[T/F]	F
-W	Word size, default if 0.	[Integer]	0
-X	X dropoff value for gapped alignment (in bits).	[Integer]	15
-Y	Effective length of the search space (use 0 for the real size).	[Real]	0
-Z	X dropoff value for final gapped alignment (in bits).	[Integer]	25

PSI-BLAST

Position-Specific Iterated BLAST (PSI-BLAST) is an iterative search in which sequences found in one round of searching are used to build a score model for the next round of searching. For details, see the "References" section at the end of this chapter.

An example of PSI-BLAST command-line entry:

```
blastpgp -i queryfilename -B alignmentfilename -j 2 -d databasefilename
```

where `-j 2` indicates to search for 2 rounds.

Most BLAST options can be used with PSI-BLAST. However, there are several *blastpgp* parameters specifically for PSI-BLAST. The following table summarizes these PSI-BLAST options.

Option	Definition	Type	Default
-c	"Constant" used in the pseudocount formula.	[Integer]	10
-h	e-value threshold for including sequences in the score matrix model.	[Real]	0.001
-j	Maximum number of rounds.	[Integer]	1
-B	Provides a way to jump start PSI-BLAST from a master-slave multiple alignment computed outside PSI-BLAST.	[File In]	
-C	Stores the query and frequency count ratio matrix in a file.	[File Out]	
-Q	Output File for PSI-BLAST Matrix in ASCII.	[File Out]	
-R	Restarts from a file stored previously.	[File In]	
-T	Produce HTML output [T/F].	[T/F]	F

PHI-BLAST

PHI-BLAST (Pattern-Hit Initiated BLAST) is a search program that combines matching of regular expressions with local alignments surrounding the match.

An example of a PHI-BLAST command-line entry:

```
blastpgp -i queryfilename -k patternfilename -p patseedp
```

where `patseedp` indicates the mode of usage.

Most BLAST options can be used with PHI-BLAST. However, the use of the `-g F` option is forbidden because PHI-BLAST requires gapped alignments.

The syntax for patterns in PHI-BLAST follows the conventions of PROSITE. All PROSITE codes are allowed, but only the ID, PA, and HI codes are relevant to PHI-BLAST.

Here is an example of a pattern:

```
ID  ER_TARGET; PATTERN.
PA  [KRHQSA]-[DENQ]-E-L>.
HI (19 22)
HI (201 204)
```

bl2seq

bl2seq (BLAST 2 Sequences) allows for the alignment of two given sequences.

An example of a *bl2seq* command-line entry:

```
bl2seq -p programname -i firstfilename -j secondfilename -o
  <outputfilenanme>
```

The following table summarizes the *bl2seq* options.

Option	Definition	Type	Default
-a	SeqAnnot output file.	[File Out]	
-d	Theoretical database size (0 is real size).	[Integer]	0
-e	Expectation value (E).	[Real]	10.0
-g	Gapped.	[T/F]	T
-i	First sequence.	[File In]	
-j	Second sequence.	[File In]	
-m	Use MegaBLAST for search.	[T/F]	F
-o	Alignment output file.	[File Out]	stdout
-p	Program name: *blastp*, *blastn*, *blastx*, *tblastn*, *tblastx*. For *blastx*, the first sequence should be nucleotide; for *tblastn*, the 2nd sequence sequence should be nucleotide.	[String]	
-q	Penalty for a nucleotide mismatch (*blastn* only).	[Integer]	-3
-r	Reward for a nucleotide match (*blastn* only).	[Integer]	1
-t	Length of the largest intron allowed in *tblastn* for linking HSPs.	[Integer]	0
-A	Input sequences in the form of <accession.version>.	[T/F]	F
-D	Output format: 0 = traditional; 1 = tabulated.	[Integer]	0
-E	Cost to extend a gap (0 invokes default behavior).	[Integer]	0
-F	Filter query sequence (DUST with *blastn*, SEG with others)	[String]	T

Option	Definition	Type	Default
-G	Cost to open a gap (0 invokes default behavior).	[Integer]	0
-I	Location on first sequence.	[String]	
-J	Location on second sequence.	[String]	
-M	Matrix.	[String]	BLOSUM62
-S	Query strands to search against database (*blastn* only). 3 is both, 1 is top, 2 is bottom.	[Integer]	3
-T	Produce HTML output.	[T/F]	F
-U	Use lowercase filtering for the query sequence.	[T/F]	F
-W	Wordsize (0 invokes default behavior).	[Integer]	0
-X	X dropoff value for gapped alignment (in bits) (0 invokes default behavior).	[Integer]	0
-Y	Effective length of the search space (use 0 for the real size).	[Real]	0

References

Altschul, S.F., W. Gish, W. Miller, E. W. Myers, and D. J. Lipman. 1990. Basic local alignment search tool. *J. Mol. Biol.* 215:403–410.

Altschul, S.F., T. L. Madden, A. A. Schäffer, J. Zhang, Z. Zhang, W. Miller, and D. J. Lipman. 1997. Gapped BLAST and PSI-BLAST: a new generation of protein database search programs. *Nucleic Acids Research* 25:3389–3402.

Gish, W., and D. J. States. 1993. Identification of protein coding regions by database similarity search. *Nature Genet.* 3:266–272.

Main page
: *http://www.ncbi.nlm.nih.gov/BLAST/*

Information guide
: *http://www.ncbi.nlm.nih.gov/Education/BLASTinfo/information3.html*

Download
: *ftp://ftp.ncbi.nih.gov/blast/executables/*

8
BLAT

BLAT (BLAST-Like Alignment Tool) is a very fast sequence alignment tool similar to BLAST. It's relatively new compared to BLAST, but is becoming very popular. We like it a lot. BLAT is more accurate and can be hundreds of times faster than BLAST. BLAT's speed comes from its runtime indexing of all nonoverlapping subsequences of given lengths. This index is small enough to fit into computer memory and is typically computed only once for each genome assembly. Jim Kent developed BLAT specifically to help with genome assembly while working on the human genome. For details see the "References" section at the end of this chapter. We're using Version 16 of BLAT.

An example of a BLAT command-line entry:

```
blat database query [-ooc=11.ooc] output.psl
```

where:

- *database* is a *.fa* file, a *.nib* file, or a list of *.fa* or *.nib* files.
- *query* is a *.fa*, *.nib*, or list of *.fa* or *.nib* files.
- `-ooc=11.ooc` tells the program to load over-occurring 11-mers from an external file. This will increase the speed by a factor of 40 in many cases, but is not required.
- `output.psl` is where to put the output.

Command-Line Options

Table 8-1 summarizes the BLAT options.

Table 8-1. BLAT options

Option	Definition	Default
-dots=N	Output dot every N sequences to show the program's progress.	
-makeOoc=N.ooc	Make overused tile file.	
-mask=type	Mask out repeats. Alignments won't be started in masked region but may extend through it in nucleotide searches. Masked areas are ignored entirely in protein or translated searches. Types are: lower = Mask out lowercased sequence. upper = Mask out uppercased sequence. out = Mask according to database.out RepeatMasker *.out* file. file.out = Mask database according to RepeatMasker *file.out*.	
-maxGap=N	Sets the size of maximum gap between tiles in a clump. Usually set from 0 to 3. Only relevant for minMatch > 1.	2
-minIdentity=N	Sets minimum sequence identity (in percent).	90 (nucleotide) 25 (protein) 25 (translated)
-minMatch=N	Sets the number of tile matches. Usually set from 2 to 4.	2 (nucleotide) 1 (protein)
-minScore=N	Sets minimum score. This is twice the matches minus the mismatches minus some sort of gap penalty.	30
-minRepDivergence=NN	Minimum percent divergence of repeats to allow them to be unmasked. Only relevant for masking using RepeatMasker *.out* files.	15
-noHead	Suppress .psl header (so it's just a tab-separated file).	
-noTrimA	Don't trim trailing poly-A.	
-oneOff=N	If set to 1, this allows one mismatch in tile and still triggers an alignment.	0
-ooc=N.ooc	Use overused tile file N.ooc. N should correspond to the tileSize.	
-out=type	Controls output file format. Type is one of: psl = Tab-separated format without actual sequence. pslx = Tab-separated format with sequence. axt = *blastz*-associated axt format. maf = *multiz*-associated maf format. wublast = similar to wublast format. blast = similar to NCBI blast format.	psl
-prot	Synonymous to -d=prot -q=prot.	
-qMask=type	Mask out repeats in query sequence. Similar to -mask, but for query rather than target sequences.	
-q=type	Query type. Type is one of: dna = DNA sequence. rna = RNA sequence. prot = protein sequence. dnax = DNA sequence translated in six frames to protein. rnax = DNA sequence translated in three frames to protein.	dna

Table 8-1. BLAT options (continued)

Option	Definition	Default
-repMatch=N	Sets the number of repetitions of a tile allowed before it is marked as overused. Typically this is: 256 for tileSize 12. 1024 for tile size 11. 4096 for tile size 10. Typically comes into play only with makeOoc.	1024
-t=type	Database type. Type is one of: dna = DNA sequence. prot = protein sequence. dnax = DNA sequence translated in six frames to protein.	dna
-tileSize=N	Sets the size of match that triggers an alignment. Usually between 8 and 12.	11 (DNA) 5 (protein)
-trimHardA	Removes poly-A tail from qSize and alignments in psl output.	
-trimT	Trims leading poly-T.	

References

Kent, W. James. 2002. BLAT—The BLAST-Like Alignment Tool. *Genome Research* 12 (4):656–664.

Main page
http://genome.ucsc.edu/cgi-bin/hgBlat?command=start

User guide
http://genome.ucsc.edu/goldenPath/help/hgTracksHelp.html

Download
http://www.soe.ucsc.edu/~kent/exe/

BLAT

9
ClustalW

ClustalW is a general-purpose multiple sequence alignment program for nucleotide sequences or proteins. The alignments can be either global (whole sequences) or local (restricted to subsequences). ClustalW calculates the best match for the selected sequences, and lines them up so that the identities, similarities, and differences can be seen. For details see the "References" section at the end of this chapter. We're using Version 1.82 of ClustalW.

An example of a ClustalW command-line entry:

```
clustalw /infile=file.txt /align
```

where *file.txt* contains the FASTA-formatted sequences.

Command-Line Options

The ClustalW options are summarized in Tables 9-1 through 9-10.

Table 9-1. ClustalW verb options

Option	Definition
-align	Do full multiple alignment.
-bootstrap(=n)	Bootstrap a NJ tree (n= number of bootstraps; def. = 1000).
-convert	Output the input sequences in a different file format.
-help or -check	Outline the command-line parameters
-options	List the command-line parameters.
-tree	Calculate NJ tree.

Table 9-2. ClustalW data options

Option	Definition
-infile=file.ext	Input sequences.
-profile1=file.ext	Profiles.
-profile2=file.ext	Profiles (old alignment).

Table 9-3. ClustalW parameters—general settings

Option	Definition
-case	LOWER or UPPER (for GDE output only).
-interactive	Read command line, then enter normal interactive menus.
-negative	Protein alignment with negative values in matrix.
-outfile=	Sequence alignment file name.
-output=	GCG, GDE, PHYLIP, or PIR.
-outorder=	INPUT or ALIGNED.
-quicktree	Use FAST algorithm for the alignment guide tree.
-seqnos=	OFF or ON (for ClustalW output only).

Table 9-4. ClustalW parameters—fast pairwise alignments

Option	Definition
-ktuple=n	Word size.
-pairgap=n	Gap penalty.
-score	PERCENT or ABSOLUTE.
-topdiags=n	Number of best diags.
-window=n	Window around best diags.

Table 9-5. ClustalW parameters—slow pairwise alignments

Option	Definition
-pwdnamatrix=	DNA weight matrix=IUB, ClustalW, or filename.
-pwgapopen=f	Gap opening penalty.
-pwgapext=f	Gap extension penalty.
-pwmatrix=	Protein weight matrix=BLOSUM, PAM, GONNET, ID or filename.

Table 9-6. ClustalW parameters—multiple alignments

Option	Definition
-dnamatrix=	DNA weight matrix=IUB, ClustalW, or filename.
-endgaps	No end gap separation penalty.
-gapdist=n	Gap separation penalty range.
-gapext=f	Gap extension penalty.
-gapopen=f	Gap opening penalty.

Table 9-6. ClustalW parameters—multiple alignments (continued)

Option	Definition
-hgapresidues=	List hydrophilic residue.
-matrix=	Protein weight matrix=BLOSUM, PAM, GONNET, ID, or filename.
-maxdiv=n	Percentage identity for delay.
-newtree=	File for new guide tree.
-nohgap	Hydrophilic gaps off.
-nopgap	Residue-specific gaps off.
-transweight	Transitions weighted.
-type=	PROTEIN or DNA.
-usetree=	File for old guide tree.

Table 9-7. ClustalW parameters—profile alignments

Option	Definition
-newtree1=	File for new guide tree for profile1.
-newtree2=	File for new guide tree for profile2.
-profile	Merge two alignments by profile alignment.
-usetree1=	File for old guide tree for profile1.
-usetree2=	File for old guide tree for profile2.

Table 9-8. ClustalW parameters—sequence to profile alignments

Option	Definition
-newtree=	File for new guide tree.
-sequences	Sequentially add profile2 sequences to profile1 alignment.
-usetree=	File for old guide tree.

Table 9-9. ClustalW parameters—structure alignments

Option	Definition
-helixendin=n	Number of residues inside helix to be treated as terminal.
-helixgap=n	Gap penalty for helix core residues.
-helixendout=n	Number of residues outside helix to be treated as terminal.
-loopgap=n	Gap penalty for loop regions.
-nosecstr1	Do not use secondary structure-gap penalty mask for profile 1.
-nosecstr2	Do not use secondary structure-gap penalty mask for profile 2.
-secstrout=	STRUCTURE or MASK or BOTH or NONE output in alignment file.
-strandgap=n	Gap penalty for strand core residues.
-strandendin=n	Number of residues inside strand to be treated as terminal.
-strandendout=n	Number of residues outside strand to be treated as terminal.
-terminalgap=n	Gap penalty for structure termini.

Table 9-10. ClustalW parameters—trees

Option	Definition
-kimura	Use Kimura's correction.
-outputtree=	nj OR phylip OR dist.
-seed=n	Seed number for bootstraps.
-tossgaps	Ignore positions with gaps

References

Higgins, D., J. Thompson, T. Gibson, J. D. Thompson, D. G. Higgins, T. J. Gibson. 1994. ClustalW: improving the sensitivity of progressive multiple sequence alignment through sequence weighting, position-specific gap penalties, and weight matrix choice. *Nucleic Acids Research* 22:4673–4680.

Main page
: *http://www.ebi.ac.uk/clustalw/*

User help
: *http://www.ebi.ac.uk/clustalw/clustalw_frame.html*

Download
: *ftp://ftp.ebi.ac.uk/pub/software/unix/clustalw/*
: *ftp://ftp.ebi.ac.uk/pub/software/dos/clustalw/*

10

HMMER

HMMER is a collection of programs that create a hidden Markov model (HMM) of a sequence family which can be utilized as a query against a sequence database to identify (and/or align) additional homologs of the sequence family. HMMER was developed by Sean Eddy at Washington University. For details, see the "References" section at the end of this chapter.

Each program and their respective options are listed below. We're using Version 2.2 of HMMER.

hmmalign

Align sequences to an HMM profile.

```
hmmalign [options] hmmfile seqfile
```

Options

-h
Print brief help; includes version number and summary of all options, including expert options.

-m
Include in the alignment only those symbols aligned to match states. Do not show symbols assigned to insert states.

-o f
Save alignment to file *f* instead of to standard output.

-q
Quiet; suppress all output except the alignment itself.

hmmbuild

Build a profile HMM from an alignment.

```
hmmbuild [options] hmmfile alignfile
```

Options

-f
Configure the model for finding multiple domains per sequence, where each domain can be a local (fragmentary) alignment.

-g
Configure the model for finding a single global alignment to a target sequence.

-h Print brief help.

-n s Name this HMM *s*.

-o f Re-save the starting alignment to *f*, in Stockholm format.

-s Configure the model for finding a single local alignment per target sequence.

-A Append this model to an existing hmmfile rather than creating hmmfile.

-F Force overwriting of an existing hmmfile.

---

hmmcalibrate

Calibrate HMM search statistics.

```
hmmcalibrate [options] hmmfile
```

Options

-h Print brief help.

---

hmmconvert

Convert between profile HMM file formats.

```
hmmconvert [options] oldhmmfile newhmmfile
```

Options

-a Convert to HMMER 2 ASCII file.

-b Convert to HMMER 2 binary file.

-h Print brief help.

-p Convert to GCG profile .prf format.

-A Append mode; append to newhmmfile rather than creating a new file.

-F Force

-P Convert the HMM to Compugen XSW extended profile format.

---

hmmemit

Generate sequences from a profile HMM.

```
hmmemit [options] hmmfile
```

Options

-a Write the generated sequences in an aligned format (SELEX) rather than FASTA.

-c Predict a single majority-rule consensus sequence instead of sampling sequences from the HMM's probability distribution.

-h Print brief help.

-n n Generate *n* sequences. Default is 10.

-o f Save the synthetic sequences to file *f* rather than writing them to stdout.

-q Quiet; suppress all output except for the sequences themselves.

hmmfetch

Retrieve an HMM from an HMM database.

```
hmmfetch [options] database name
```

Options

-h Print brief help.

hmmindex

Create a binary SSI index for an HMM database.

```
hmmindex [options] database
```

Options

-h Print brief help.

hmmpfam

Search one or more sequences against an HMM database.

```
hmmpfam [options] hmmfile seqfile
```

Options

-h Print brief help.

-n Specify that models and sequence are nucleic acid, not protein.

-A n Limits the alignment output to the *n* best scoring domains.

-E x Set the E-value cutoff for the per-sequence ranked hit list to *x*, where *x* is a positive real number.

-T x Set the bit score cutoff for the per-sequence ranked hit list to *x*, where *x* is a real number.

-Z n Calculate the E-value scores as if we had seen a sequence database of *n* sequences.

hmmsearch

Search a sequence database with a profile HMM.

```
hmmsearch [options] hmmfile seqfile
```

Options

-h Print brief help; includes version number and summary of all options, including expert options.

-A n Limits the alignment output to the *n* best scoring domains.

-E x Set the E-value cutoff for the per-sequence ranked hit list to *x*, where *x* is a positive real number.

-T x Set the bit score cutoff for the per-sequence ranked hit list to *x*, where *x* is a real number.

-Z n Calculate the E-value scores as if we had seen a sequence database of *n* sequences.

References

Durbin, Richard, Sean Eddy, Anders Krogh, and Graeme Mitchison. 1998. *Biological Sequence Analysis: Probabilistic Models of Proteins and Nucleic Acids.* Cambridge: Cambridge University Press.

Eddy, S.R. 1998. Profile hidden Markov models. *Bioinformatics* 14:755–763.

Main page
http://hmmer.wustl.edu/

README
ftp://ftp.genetics.wustl.edu/pub/eddy/hmmer/CURRENT/00README

Download
ftp://ftp.genetics.wustl.edu/pub/eddy/hmmer/2.2g/hmmer-2.2g.tar.gz

11
MEME/MAST

The MEME/MAST system allows you to:

1. Discover motifs (highly conserved regions) in groups of related DNA or protein sequences using MEME.
2. Search sequence databases using motifs using MAST.

MEME and MAST were developed by Timothy Bailey, Charles Elkan, and Bill Grundy at the UCSD Computer Science and Engineering department with input from Michael Gribskov at the San Diego Supercomputer Center.

We're using Version 3.0.4 of MEME/MAST.

MEME

MEME (Multiple EM for Motif Elicitation) is a tool for discovering motifs in a group of related DNA or protein sequences.

A motif is a sequence pattern that occurs repeatedly in a group of related protein or DNA sequences. MEME represents motifs as position-dependent letter-probability matrices which describe the probability of each possible letter at each position in the pattern. Individual MEME motifs do not contain gaps. Patterns with variable-length gaps are split by MEME into two or more separate motifs.

MEME takes as input a group of DNA or protein sequences (the training set) and outputs as many motifs as requested. MEME uses statistical modeling techniques to automatically choose the best width, number of occurrences, and description for each motif. For details, see the "References" section at the end of this chapter.

Examples

The following examples use data files provided in this release of MEME. MEME writes its output to standard output, so you will want to redirect it to a file in order for use with MAST.

A simple DNA example:

```
meme crp0.s -dna -mod oops -pal > ex1.html
```

MEME looks for a single motif in the file *crp0.s* which contains DNA sequences in FASTA format. The OOPS model is used so MEME assumes that every sequence contains exactly one occurrence of the motif. The palindrome switch is given so the motif model (PSPM) is converted into a palindrome by combining corresponding frequency columns. MEME automatically chooses the best width for the motif in this example since no width was specified.

Searching for motifs on both DNA strands:

```
meme crp0.s -dna -mod oops -revcomp > ex2.html
```

This is like the previous example except that the `-revcomp` switch tells MEME to consider both DNA strands, and the `-pal` switch is absent so the palindrome conversion is omitted. When DNA uses both DNA strands, motif occurrences on the two strands may not overlap. That is, any position in the sequence given in the training set may be contained in an occurrence of a motif on the positive strand or the negative strand, but not both.

A fast DNA example:

```
meme crp0.s -dna -mod oops -revcomp -w 20 > ex3.html
```

This example differs from the first example in that MEME is told to only consider motifs of width 20. This causes MEME to execute about 10 times faster. The `-w` switch can also be used with protein datasets if the width of the motifs are known in advance.

Using a higher-order background model:

```
meme INO_up800.s -dna -mod tcm -revcomp -bfile yeast.nc.6.freq > ex4.html
```

In this example we use `-mod tcm` and `-bfile` *yeast.nc.6.freq*. This specifies that:

- The motif may have any number of occurrences in each sequence.
- The Markov model specified in *yeast.nc.6.freq* is used as the background model. This file contains a fifth-order Markov model for the non-coding regions in the yeast genome.

Using a higher-order background model can often result in more sensitive detection of motifs. This is because the background model more accurately models non-motif sequence, allowing MEME to discriminate against it and find the true motifs.

A simple protein example:

```
meme lipocalin.s -mod oops -maxw 20 -nmotifs 2 > ex5.html
```

The -dna switch is absent, so MEME assumes the file *lipocalin.s* contains protein sequences. MEME searches for two motifs each of width less than or equal to 20. (Specifying -maxw 20 makes MEME run faster, since it does not have to consider motifs longer than 20.) Each motif is assumed to occur in each of the sequences because the OOPS model is specified.

Another simple protein example:

```
meme farntrans5.s -mod tcm -maxw 40 -maxsites 50 > ex6.html
```

MEME searches for a motif of width up to 40, with up to 50 occurrences in the entire training set. The TCM sequence model is specified, which allows each motif to have any number of occurrences in each sequence. This dataset contains motifs with multiple repeats of motifs in each sequence. This example is fairly time consuming due to the fact that the time required to initialize the motif probability tables is proportional to maxw multiplied by maxsites. By default, MEME only looks for motifs up to 29 letters wide with a maximum total of number of occurrences equal to twice the number of sequences or 30, whichever is less.

A much faster protein example:

```
meme farntrans5.s -mod tcm -w 10 -maxsites 30 -nmotifs 3 > ex7.html
```

This time MEME is constrained to search for three motifs of width exactly ten. The effect is to break up the long motif found in the previous example. The -w switch forces motifs to be *exactly* ten letters wide. This example is much faster because, since only one width is considered, the time to build the motif probability tables is only proportional to maxsites.

Splitting the sites into three:

```
meme farntrans5.s -mod tcm -maxw 12 -nsites 24 -nmotifs 3 > ex8.html
```

This forces each motif to have exactly 24 occurrences, and be up to 12 letters wide.

A larger protein example with E-value cutoff:

```
meme adh.s -mod zoops -nmotifs 20 -evt 0.01 > ex9.html
```

In this example, MEME looks for up to 20 motifs, but stops when a motif is found with E-value greater than 0.01. Motifs with large E-values are likely to be statistical artifacts rather than biologically significant.

Command-Line Options

Usage for MEME is:

```
meme dataset optionalarguments
```

where *dataset* is a file containing sequences in FASTA format.

Table 11-1 summarizes the command-line options for MEME.

Table 11-1. MEME options

Option	Definition
[-h]	Print this message.
[-dna]	Sequences use DNA alphabet.
[-protein]	Sequences use protein alphabet.
[-mod oops|zoops|tcm]	Distribution of motifs.
[-nmotifs *nmotifs*]	Maximum number of motifs to find.
[-evt *ev*]	Stop if motif E-value greater than *evt*.
[-nsites *sites*]	Number of sites for each motif.
[-minsites *minsites*]	Minimum number of sites for each motif.
[-maxsites *maxsites*]	Maximum number of sites for each motif.
[-wnsites *wnsites*]	Weight on expected number of sites.
[-w *w*]	Motif width.
[-minw *minw*]	Minumum motif width.
[-maxw *maxw*]	Maximum motif width.
[-nomatrim]	Do not adjust motif width using multiple alignment.
[-wg *wg*]	Gap opening cost for multiple alignments.
[-ws *ws*]	Gap extension cost for multiple alignments.
[-noendgaps]	Do not count end gaps in multiple alignments.
[-bfile *bfile*]	Name of background Markov model file.
[-revcomp]	Allow sites on + or - DNA strands.
[-pal]	Force palindromes (requires -dna).
[-maxiter *maxiter*]	Maximum EM iterations to run.
[-distance *distance*]	EM convergence criterion.
[-prior dirichlet|dmix|mega|megap|addone]	Type of prior to use.
[-b *b*]	Strength of the prior.
[-plib *plib*]	Name of Dirichlet prior file.
[-spfuzz *spfuzz*]	Fuzziness of sequence to theta mapping.
[-spmap uni|pam]	Starting point seq to theta mapping type.
[-cons *cons*]	Consensus sequence to start EM from.
[-text]	Output in text format (default is HTML).
[-print_fasta]	Print sites in FASTA format (default BLOCKS).
[-maxsize *maxsize*]	Maximum dataset size in characters.
[-nostatus]	Do not print progress reports to terminal.
[-p *np*]	Use parallel version with *np* processors.
[-time *t*]	Quit before *t* CPU seconds consumed.
[-sf *sf*]	Print *sf* as name of sequence file.

MAST

MAST (Motif Alignment and Search Tool) is a tool for searching biological sequence databases for sequences that contain one or more of a group of known motifs.

A motif is a sequence pattern that occurs repeatedly in a group of related protein or DNA sequences. Motifs are represented as position-dependent scoring matrices that describe the score of each possible letter at each position in the pattern. Individual motifs may not contain gaps. Patterns with variable-length gaps must be split into two or more separate motifs before being submitted as input to MAST.

MAST takes as input a file containing the descriptions of one or more motifs and searches a sequence database that you select for sequences that match the motifs. The motif file can be the output of the MEME motif discovery tool or any file in the appropriate format. For details, see the "References" section at the end of this chapter.

Examples

The following examples assume that file *meme.results* is the output of a MEME run containing at least 3 motifs and file *SwissProt* is a copy of the SWISS-PROT database on your local disk. DNA_DB is a copy of a DNA database on your local disk.

Annotate the training set:

```
mast meme.results
```

Find sequences matching the motif and annotate them in the SWISS-PROT database:

```
mast meme.results -d SwissProt
```

Show sequences with weaker combined matches to motifs:

```
mast meme.results -d SwissProt -ev 200
```

Indicate weaker matches to single motifs in the annotation so that sequences with weak matches to the motifs (but perhaps with the "correct" order and spacing) can be seen:

```
mast meme.results -d SwissProt -w
```

Include a nominal order and spacing of the first three motifs in the calculation of the sequence p-values to increase the sensitivity of the search for matching sequences:

```
mast meme.results -d SwissProt -diag "9-[2]-61-[1]-62-[3]-91"
```

Use only the first and third motifs in the search:

```
mast meme.results -d SwissProt -m 1 -m 3
```

Use only the first two motifs in the search:

```
mast meme.results -d SwissProt -c 2
```

Search DNA sequences using protein motifs, adjusting p-values and E-values for each sequence by that sequence's composition:

```
mast meme.results -d DNA_DB -dna -comp
```

Command-Line Options

Usage for MAST is the following.

```
mast mfile optionalarguments ...
```

where *mfile* is a file containing motifs to use. This may be a MEME output file, or a file with the format described in the MAST manpage at *http://meme.sdsc.edu/meme/website/meme-download.html*.

Table 11-2 summarizes the command-line options for MAST.

Table 11-2. MAST options

Option	Definition
mfile	File containing motifs to use; may be a MEME output file or a file with a supported format.
[*database*]	Database containing motifs to use.
[-d *database*]	Database to search with motifs.
[-stdin]	Read database from standard input; default reads database specified inside *mfile*.
[-c *count*]	Only use the first *count* motifs.
[-a *alphabet*]	*mfile* is assumed to contain motifs in the format output by bin/make_logodds and *alphabet* is their alphabet; -d *database* or -stdin must be specified when this option is used.
[-stdout]	Print output to standard output instead of a file.
[-text]	Output in text (ASCII) format; default is hypertext (HTML) format.
[-sep]	Score reverse complement DNA strand as a separate sequence.
[-norc]	Do not score reverse complement DNA strand.
[-dna]	Translate DNA sequences to protein.
[-comp]	Adjust p-values and E-values for sequence composition.
[-rank *rank*]	Print results starting with *rank* best; default is1.
[-smax *smax*]	Print results for no more than *smax* sequences; default is all.
[-ev *ev*]	Print results for sequences with E-value *ev*; default is 10.
[-mt *mt*]	Show motif matches with p-value *mt* ; default is 0.0001.
[-w]	Show weak matches (mt<p-value<mt*10) in angle brackets.
[-bfile *bfile*]	Read background frequencies from *bfile*.
[-seqp]	Use SEQUENCE p-values for motif thresholds (default: use POSITION p-values).
[-mf *mf*]	Print *mf* as motif file name.
[-df *df*]	Print *df* as database name.
[-minseqs *minseqs*]	Lower bound on number of sequences in db.
[-mev *mev*]+	Use only motifs with E-values less than *mev*.
[-m *m*]+	Use only motif(s) number *m* (overrides -mev).
[-diag *diag*]	Nominal order and spacing of motifs.
[-best]	Include only the best motif in diagrams.
[-remcorr]	Remove highly correlated motifs from query.
[-brief]	Brief output—do not print documentation.
[-b]	Print only sections I and II.
[-nostatus]	Do not print progress report.

References

Bailey, Timothy L., and Charles Elkan. 1994. Fitting a mixture model by expectation maximization to discover motifs in biopolymers. *Proceedings of the Second International Conference on Intelligent Systems for Molecular Biology* 28–36. Menlo Park: AAAI Press.

Bailey, Timothy L., and Michael Gribskov. 1998. Combining evidence using p-values: application to sequence homology searches. *Bioinformatics* 14:48–54.

Main page
http://meme.sdsc.edu/meme/website/intro.html

Manpages
http://meme.sdsc.edu/meme/website/meme-download.html

Download
ftp://ftp.sdsc.edu/pub/sdsc/biology/meme

12
EMBOSS

EMBOSS (European Molecular Biology Open Software Suite) is an open source package of sequence analysis tools. This software covers a wide range of functionality and can handle data in a variety of formats. Extensive libraries are provided with the package, allowing users to develop and release their own software. EMBOSS also integrates a range of currently available packages and tools for sequence analysis, such as BLAST and ClustalW. A Java API (Jemboss) is also available.

EMBOSS contains around 150 programs (applications). These are just some of the areas covered:

- Sequence alignment.
- Rapid database searching with sequence patterns.
- Protein motif identification, including domain analysis.
- Nucleotide sequence pattern analysis, for example to identify CpG islands or repeats.
- Codon usage analysis for small genomes.
- Rapid identification of sequence patterns in large scale sequence sets.
- Presentation tools for publication.

...and much more.

For details, see the "References" section at the end of this chapter.

We're using Version 2.5.0 of EMBOSS.

Common Themes

Many EMBOSS programs have functionality in common. They all understand the same sorts of sequence addresses, sequence formats, output formats, and feature formats. The following sections describe some common themes in EMBOSS.

Uniform Sequence Address

The Uniform Sequence Address (USA) is a standard sequence naming used by all EMBOSS applications.

The USA syntax is one of:

- "*format*::*file*"
- "*format*::*file*:*entry*"
- "*dbname*:*entry*"
- "*@listfile*" (a file of filenames)

The "::" and ":" syntax is to allow, for example, "embl" and "pir" to be both database names and sequence formats. In addition, EMBOSS allows the command line to separately define the format and the entry name so that only the filename is required.

The "file" and "dbname" forms of USA may have "format::" in front of them, but because a database is aware of the format, this structure is redundant and not recommended.

Any USA may optionally take this subsequence specifier after the main body of the USA, either in the form "[*start* : *end*]" or "[*start* : *end* : r]", where *start* and *end* are the required start and end positions. Negative positions count from the end of the sequence. Use of this USA subsequence specifier is equivalent to using the `-sbegin`, `-send`, or `-sreverse` command-line qualifiers.

Table 12-1 contains some USA examples.

Table 12-1. Emboss Uniform Sequence Address (USA) examples

Type	Example	Comments
filename	*xxx.seq*	A sequence file *xxx.seq* in any format.
format::*filename*	fasta::*xxx.seq*	A sequence file *xxx.seq* in FASTA format.
db:*IDname*	embl:paamir	EMBL entry PAAMIR, using whatever access method is defined locally for the EMBL database.
db:*AccessionNumber*	embl:X13776	EMBL entry X13776, using whatever access method is defined locally for the EMBL database. Search by accession number and entry name. X13776 is the accession number in this case.
db-acc:*AccessionNumber*	embl-acc:X13776	EMBL entry X13776, using whatever access method is defined locally for the EMBL database. Search by accession number only.
db-id:*IDname*	embl-id:paamir	EMBL entry PAAMIR, using whatever access method is defined locally for the EMBL database. Search by ID only.
db-searchfield:*word*	embl-des:lectin	EMBL entries containing the word "lectin" in the Description line.
db-searchfield:*wcardword*	embl-org:*human*	EMBL entries containing the wildcarded word "human" in the Organism fields.
db:*wildcard-ID*	embl:paami*	EMBL entries PAAMIB, PAAMIE and so on, usually in alphabetical order, using whatever access method is defined locally for the EMBL database.

Table 12-1. Emboss Uniform Sequence Address (USA) examples (continued)

Type	Example	Comments
db or *db*:*	embl or EMBL:*	All sequences in the EMBL database.
`@listfile`	@mylist	Reads file *mylist* and uses each line as a separate USA. List files can contain references to other lists files or any other standard USA.
`list:listfile`	list:mylist	Same as `@mylist`.
`programparameters` |	getz -e [embl-id:paamir] |	The pipe character "|" causes EMBOSS to fire up *getz* (the SRS sequence retrieval program) to extract entry PAAMIR from EMBL in EMBL format. Any application or script which writes one or more sequences to `stdout` can be used in this way.
asis::`sequence`	asis::atacgcagttatctgaccat	So far, the shortest USA we could invent. In "asis" format the name is the sequence, so no file needs to be opened. This is a special case. It was intended as a joke, but could be quite useful for generating command lines.

EMBOSS

Sequence Formats

You can specify the format to use on input by giving the format name with two colons before the file holding your sequences. For example:

```
embl::myfile.seq
```

The format is not required. When reading in a sequence, EMBOSS will guess the sequence format by trying all known formats until one succeeds.

When writing out a sequence, EMBOSS will use FASTA format by default. You can specify another format to use, for example:

```
gcg::myresults.seq
```

Input sequence formats

To date, the sequence formats in Table 12-2 are accepted as input. By default (i.e., no format is explicitly specified), EMBOSS tries each format in turn until one succeeds.

Table 12-2. EMBOSS input sequence formats

Input format	Comments
abi	ABI trace file format. This is the format of file produced by ABI sequencing machines. It contains the *trace data*, i.e., the probabilities of the 4 bases along the sequencing run, together with the sequence, as deduced from that data. The sequence information is what is normally read in and used by EMBOSS programs, although the trace data is available and may be utilized by some specialized EMBOSS programs. The code for this is heavily based on David Mathog's Fortran library with a description of ABI trace file format (*abi.txt*): *ftp://saf.bio.caltech.edu/pub/software/molbio/abitools.zip*.
acedb	ACeDB format.
clustal aln	ClustalW ALN (multiple alignment) format.
codata	CODATA format.

Table 12-2. EMBOSS input sequence formats (continued)

Input format	Comments

dbid	Odd FASTA format with Database name first, folowed by ID name and an optional accession number, e.g.:
	`>database name description`
	or
	`>database name accession description embl`

em	EMBL entry format, or at least a minimal subset of the fields. The Staden package and others use EMBL or similar formats for sequence data.

pearson	FASTA format with an optional accession number after the sequence identifier, e.g.:
	`>name description`
	or
	`>name accession description`
	and with an optional database name in GCG style FASTA format included as part of the sequence identifier, e.g.:
	`>database:name accession description`

gcg gcg8	GCG 9.x and 10.x format with the format and sequence type identified on the first line of the file. GCG 8.x format where anything up to the first line containing ".." is considered as heading, and the remainder is sequence data.
genbank gb ddbj	GENBANK entry format, or at least a minimal subset of the fields.
gff	GFF format.
hennig86	Hennig86 format.
ig	IntelliGenetics format.
jackknifer	Jackknifer format.
jackknifernon	Jackknifernon format.
nbrf pir	NBRF (PIR) format, as used in the PIR database sequence files.
nexus paup	Nexus/PAUP format.
nexusnon paupnon	Nexusnon/PAUPnon format.
treecon	Treecon format.
mega	Mega format.
meganon	Meganon format.
msf	Wisconsin Package GCG's MSF multiple sequence format.
ncbi	FASTA format with optional accession number and database name in NCBI style included as part of the sequence identifier, e.g.: `>database|accession|id description` (and other variants on this theme!)
pfam stockholm	Pfam format.
phylip	PHYLIP interleaved multiple alignment format.
selex	SELEX format is used by Sean Eddy's HMMER package. It can store RNA secondary structure as part of the sequence annotation.
staden experiment	The experiment file format used by the *gap* program in the Staden package, where the sequence identifier is optional and the remainer is plain text. Some alternative nucleotide ambiguity codes are used and must be converted.

Table 12-2. EMBOSS input sequence formats (continued)

Input format	Comments
strider	DNA Strider format.
swissprot swiss sw	SWISS-PROT entry format, or at least a minimal subset of the fields.
text plain	Plain text. This is the format with no format. The whole of the file is read in as a sequence. No attempt is made to parse the file contents in any way. Anything is acceptable in this format. This means that any character will be included in the sequence, even digits and punctuation. Use this format only when you are sure that the input sequence file is correct and contains only what you want to be considered as your sequence.
raw	Similar to text or plain format. However, raw removes any whitespace or digits, accepts only alphabetic characters, and rejects anything else. This format is safer than plain format. Digits, spaces, and TAB characters are removed and ignored. If a sequence contains other non-alphabetic characters (e.g., punctuation characters), it is rejected as erroneous.
asis	Not a sequence format , but a quick way of entering a sequence on the command line. It is included here for completeness. In "asis" format, the actual sequence appears where a filename would normally be given.
asis::atacgcagttatctgacc	In "asis" format the name is the sequence, so no file needs to be opened. This is a special case. It was intended as a joke, but could be quite useful for generating command lines.

EMBOSS

Output sequence formats

To date, the sequence formats in Table 12-3 are available as output. Some sequence formats can hold multiple sequences in one file; these are marked as multiple in the table. Formats such as GCG, plain, and staden can hold only one sequence per file and are marked as single.

Table 12-3. EMBOSS input sequence formats

Output format	Single/multiple	Comments
gcg gcg8	single	Wisconsin Package GCG 9.x and 10.x format with the sequence type on the first line of the file. GCG 8.x format where anything up to the first line containing ".." is considered as heading, and the remainder is sequence data.
embl em	multiple	EMBL entry format with available fields filled in and others with no information omitted. The EMBOSS command line allows missing data such as accession numbers to be provided if they are not obtainable from the input sequence.
swiss sw	multiple	SwisProt entry format with available fields filled in and others with no information omitted. The EMBOSS command line allows missing data such as accession numbers to be provided if they are not obtainable from the input sequence.
fasta pearson	multiple	Standard Pearson FASTA format, but with the accession number included after the identifier if available.
ncbi	multiple	NCBI style FASTA format with the database name, entry name and accession number separated by pipe ("|") characters.
nbrf pir	multiple	NBRF (PIR) format, as used in the PIR database sequence files.
genbank gb	multiple	GENBANK entry format with available fields filled in and others with no information omitted. The EMBOSS command line allows missing data such as accession numbers to be provided if they are not obtainable from the input sequence.

Table 12-3. EMBOSS input sequence formats (continued)

Output format	Single/multiple	Comments
gff	multiple	GFF format.
ig	multiple	IntelliGenetics format, as used by the IntelliGenetics package.
codata	multiple	CODATA format.
stride	multiple	DNA strider format.
acedb	multiple	ACeDB format.
staden experiment	single	The experiment file format used by the *gap* program in the Staden package. Some alternative nucleotide ambiguity codes are used and are converted.
text plain raw	single	Plain sequence, no annotation or heading.
fitch	multiple	Fitch format.
msf	multiple	Wisconsin Package GCG's MSF multiple sequence format.
clustal aln	multiple	Clustal multiple sequence format.
selex	multiple	SELEX format.
phylip	multiple	PHYLIP interleaved format.
phylip3	multiple	PHYLIP non-interleaved format that was used in Phylip version 3.2.
asn1	multiple	A subset of ASN.1 containing entry name, accession number, description and sequence, similar to the current ASN.1 output of Readseq.
hennig86	multiple	Hennig86 format.
mega	multiple	Mega format.
meganon	multiple	Meganon format.
nexus paup	multiple	Nexus/PAUP format.
nexusnon paupnon	multiple	Nexusnon/PAUPnon format.
jackknifer	multiple	Jackknifer format.
jackknifernon	multiple	Jackknifernon format.
treecon	multiple	Treecon format.
debug	multiple	EMBOSS sequence object report for debugging showing all available fields. Not all fields will contain data—this depends very much on the input format used.

Alignment Formats

When writing out an alignment between two or more sequences, EMBOSS now uses a standard set of formats.

Multiple sequence alignment formats

Table 12-4 contains details about the current set of multiple sequence alignment formats available in EMBOSS.

Table 12-4. EMBOSS multiple sequence alignment formats

Name	Comments
unknown multiple simple	These are synonyms for simple format. This format displays the sequence names, positions and sequences, then puts the markup line underneath the sequences. When only two sequences are being aligned, the format is changed to that produced by pair.
fasta	This is the standard FASTA sequence format with gaps, where many sequences are concatenated one after the other.
msf	This is the standard MSF sequence format.
trace	This is a special verbose format for use in debugging. It is not intended for normal users.
srs	This shows the sequence ID name, the sequence position, the sequence and the sequence position for each line.

Pairwise sequence alignment formats

Table 12-5 contains details about the current set of pairwise sequence alignment formats available in EMBOSS.

Table 12-5. EMBOSS pairwise sequence alignment formats

Name	Comments
pair	This is the default format used when there are only 2 sequences. When simple format is selected but there are only 2 sequences, this format is used. The sequences have the markup line between them.
markx0	This is the standard default output format used by Bill Pearson's suite of FASTA programs.
markx1	This is an alternative output format used by Bill Pearson's suite of FASTA programs in which identities are not marked. Instead, conservative replacements are denoted by "x" and non-conservative substitutions by "X".
markx2	This is an alternative output format used by Bill Pearson's suite of FASTA programs in which the residues in the second sequence are only shown if they are different from the first.
markx3	This is an alternative output format used by Bill Pearson's suite of FASTA programs in which the aligned sequences are displayed in FASTA sequence format. These can be used to build a primitive multiple alignment.
markx10	This is an alternative output format used by Bill Pearson's suite of FASTA programs in which the aligned sequences are displayed in FASTA sequence format and the sequence length, alignment start and stop information is given in lines starting with a ";" character just after the title line for each sequence. It is intended to be easily parsed by other programs.
srspair	This is very similar in style to pair format.
score	This does not display the sequence alignment. It shows only the names of the sequences, the length of the alignment, and the score.

Feature Formats

When reading or writing features associated with a sequence, a standard set of formats is used. The feature files can either be a standard sequence format with a feature table as part of the sequence format, or the features can be held in a file without the associated sequence.

Table 12-6 contains details about the current set of feature formats available in EMBOSS.

Table 12-6. EMBOSS feature formats

Name	Comments
embl em	The format used by the EMBL nucleic database.
gff	The General Feature Format defined by the Sanger Centre.
swissprot swiss sw	The format used by the SWISS-PROT protein database. The feature table keys are also defined.
pir	The format used by the PIR protein database.
nbrf	Only available for input—the same as PIR format.

Report Formats

There are many ways in which the results of an analysis can be reported. Many EMBOSS programs are now able to output their results in a standard report format—you can change the report format used by putting `-rformat` *name* on the command line, where *name* is the name of one of the standard report formats.

Table 12-7 contains examples of *garnier* analyzing sw:100K_rat output in various report formats.

Table 12-7. EMBOSS report formats

Name	Comments
embl	Writes a report in EMBL feature table format.
genbank	Writes a report in Genbank feature table format.
gff	Writes a report in GFF feature table format.
pir	Writes a report in PIR feature table format.
swiss	Writes a report in SWISS-PROT feature table format.
trace	Of use only for debugging.
listfile	Writes out a list file with the start and end points of the motifs given by "[start:end]" after the sequence's full USA. This is useful as it is a true List File that can be read in by other EMBOSS programs using "@" or "list::" before the filename.

dbmotif	Writes a report in DbMotif format.
	Format:
	`Length = [length]`
	`Start = position [start] of sequence`
	`End = position [end] of sequence`
	`... other tags ...`
	`[sequence]`
	`[start and end numbered below sequence with '|' marks]`
	`Blank line`
	Data reported: Length, Start, End, Sequence (5 bases around feature)

Table 12-7. EMBOSS report formats (continued)

<table>
<tr><th>Name</th><th>Comments</th></tr>
<tr><td>diffseq</td><td>This format is most useful when reporting the results of two aligned sequences, as in the program diffseq. The report describes matches, usually short, between two sequences and features which overlap them.
Format:
<pre>
[Sequence 1 Name] [start]-[end] Length: [length]
Feature: first sequence feature(s)
Sequence: motif in sequence 1
Sequence: motif in sequence 2
Feature: second sequence feature(s)
[Sequence 2 Name] [start]-[end] Length: [length]
Blank line
</pre></td></tr>
<tr><td>excel</td><td>A TAB-delimited table format suitable for reading into spreadsheet programs such as Excel. Name, start, end, and score are always reported. Other tags in the report definition are added as extra columns. All values are (for now) unquoted. Missing values are reported as ".".</td></tr>
<tr><td>feattable</td><td>Writes a report in FeatTable format. The report is an EMBL feature table using only the tags in the report definition. There is no requirement for tag names to match standards for the EMBL feature table. The original EMBOSS application for this format was cpgreport.
Format:
<pre>
FT [type] [start]..[end]
FT   /[tagname]=[tagvalue]
Blank line
</pre>
Data reported: Type, Start, End</td></tr>
<tr><td>motif</td><td>Writes a report in Motif format. Based on the original output format of antigenic, helixturnhelix and sigcleave.
Format:
<pre>
(1) Score [score] length [length] at [name] [start->[end]
              * (marked at position pos)
            [sequence]
            | |
        [start]   [end]
[tagname]: tagvalue
</pre>
Data reported: Name, Start, End, Length, Score, Sequence</td></tr>
<tr><td>regions</td><td>Writes a report in Regions format. The report (unusually for the current report formats) includes the feature type.
Format:
<pre>
[type] from [start] to [end] ([length] [name]) ([tagname]:
[tagvalue], [tagname]: [tagvalue] ...)
</pre>
Data reported: Type, Start, End, Length, Name</td></tr>
<tr><td>seqtable</td><td>Writes a report in SeqTable format. This is a simple table format that includes the feature sequence. See the following "table" entry for a version without the sequence. Missing tag values are reported as "." The column width is 6, or longer if the name is longer.
Format:
<pre>
Start   End   [tagnames] Sequence
[start] [end] [tagvalues] [sequence]
</pre></td></tr>
</table>

Table 12-7. EMBOSS report formats (continued)

Name	Comments

simple	Writes a report in SRS simple format. This is a simple parsable format that does not include the feature sequence (see also SRS format) for applications where features can be large. Missing tag values are reported as ".".
	Format:
	`Feature [number]`
	`Name: [ID name]`
	`Start:  [start]`
	`End: [end]`
	`Length: [length]`
	`[tagnames:]  [tag values]`
	`Blank line`

srs	Writes a report in SRS format. This is a simple parsable format that includes the feature sequence. Missing tag values are reported as ".".
	Format:
	`Feature [number]`
	`Name: [ID name]`
	`Start:  [start]`
	`End: [end]`
	`Length: [length]`
	`Sequence: [sequence]`
	`Score: [score]`
	`[tagnames:]  [tag values]`
	`Blank line`

table	Writes a report in Table format. See previous "seqtable" entry for a version with the sequence. Missing tag values are reported as ".". The column width is 6, or longer if the name is longer.
	Format:
	`USA    Start   End   Score   [tagnames]`
	`[name] [start] [end] [score] [tagvalues]`

tagseq	Writes a report in Tagseq format. Features are marked up below the sequence. Originally developed for the *garnier* application, this format also has general uses.
	Format:
	`Sequence position written every 10 bases/residues`
	`Sequence (50 residues)`
	`tagname        ++++++++++++    +++++++++`
	`Blank line`
	If the tag value is a 1-letter code, use it in place of "+".

EMBOSS Application Groups

To aid users in finding programs of interest, the EMBOSS developers have clustered the programs into application groups. These groups are presented below.

Alignment consensus

cons *megamerger* *merger*

Alignment differences

diffseq

Alignment dot plots

dotmatcher *dotpath* *dottup* *polydot*

Alignment global

alignwrap *est2genome* *needle* *stretcher*

Alignment local

matcher *seqmatchall* *supermatcher* *water* *wordmatch*

Alignment multiple

emma *plotcon* *showalign*
infoalign *prettyplot* *tranalign*

Display

abiview *pepnet* *prettyseq* *showalign* *showseq*
cirdna *pepwheel* *remap* *showdb* *textsearch*
lindna *prettyplot* *seealso* *showfeat*

Edit

cutseq *listor* *nthseq* *splitter* *yank*
biosed *extractseq* *notseq* *skipseq* *vectorstrip*
degapseq *maskfeat* *pasteseq* *swissparse*
descseq *maskseq* *revseq* *trimest*
entret *newseq* *seqret* *trimseq*
extractfeat *noreturn* *seqretsplit* *union*

Enzyme kinetics

findkm

Feature tables

coderet *extractfeat* *maskfeat* *showfeat* *swissparse*

Information

infoalign *seealso* *textsearch* *whichdb* *wossname*
infoseq *showdb* *tfm*

Menus

emnu

Nucleic 2d structure

einverted

Nucleic codon usage

cai	*chips*	*codcmp*	*cusp*	*syco*

Nucleic composition

banana	*chaos*	*dan*	*isochore*
btwisted	*compseq*	*freak*	*wordcount*

Nucleic cpg islands

cpgplot	*cpgreport*	*geecee*	*newcpgreport*	*newcpgseek*

Nucleic gene finding

getorf	*marscan*	*plotorf*	*showorf*	*wobble*

Nucleic motifs

dreg	*fuzznuc*	*fuzztran*	*marscan*

Nucleic mutation

msbar	*shuffleseq*

Nucleic primers

eprimer3	*primersearch*	*stssearch*

Nucleic profiles

profit	*prophecy*	*prophet*

Nucleic repeats

einverted	*equicktandem*	*etandem*	*palindrome*

Nucleic restriction

recoder	*remap*	*restrict*	*silent*
redata	*restover*	*showseq*	

Nucleic transcription

tfscan

Nucleic translation

backtranseq, *coderet*, *plotorf*, *prettyseq*, *remap*, *showorf*, *showseq*, *transeq*

Phylogeny

distmat

Protein 2d structure

garnier, *helixturnhelix*, *hmoment*, *pepcoil*, *pepnet*, *pepwheel*, *tmap*

Protein 3d structure

contacts, *dichet*, *hmmgen*, *interface*, *profgen*, *psiblasts*, *scopalign*, *scoprep*, *scopreso*, *seqalign*, *seqsearch*, *seqsort*, *seqwords*, *siggen*, *sigscan*

Protein composition

backtranseq, *charge*, *checktrans*, *compseq*, *emowse*, *freak*, *iep*, *mwcontam*, *mwfilter*, *octanol*, *pepinfo*, *pepstats*, *pepwindow*, *pepwindowall*

Protein motifs

antigenic, *digest*, *fuzzpro*, *fuzztran*, *helixturnhelix*, *oddcomp*, *patmatdb*, *patmatmotifs*, *pepcoil*, *preg*, *pscan*, *sigcleave*

Protein mutation

msbar, *shuffleseq*

Protein profiles

profit, *prophecy*, *prophet*

Protein structure

seqsort

Test

histogramtest

Utilities—database creation

aaindexextract	*groups*	*pdbtosp*	*scope*	*tfextract*
cutgextract	*hetparse*	*printsextract*	*scopnr*	
domainer	*nrscope*	*prosextract*	*scopparse*	
funky	*pdbparse*	*rebaseextract*	*scopseqs*	

Utilities—database indexing

dbiblast *dbifasta* *dbiflat* *dbigcg*

Utilities—miscellaneous

embossdata *embossversion*

List of All EMBOSS Programs

Table 12-8 contains one-line descriptions of all the EMBOSS programs.

Table 12-8. EMBOSS program descriptions

Program	Description
aaindexextract	Extract data from AAINDEX.
abiview	Reads ABI file and display the trace.
alignwrap	Aligns a set of sequences to a seed alignment.
antigenic	Finds antigenic sites in proteins.
backtranseq	Back translate a protein sequence..
banana	Bending and curvature plot in B-DNA.
biosed	Replace or delete sequence sections.
btwisted	Calculates the twisting in a B-DNA sequence.
cai	CAI codon adaptation index.
chaos	Create a chaos game representation plot for a sequence.
charge	Protein charge plot.
checktrans	Reports STOP codons and ORF statistics of a protein sequence.
chips	Codon usage statistics.
cirdna	Draws circular maps of DNA constructs.
codcmp	Codon usage table comparison
coderet	Extract CDS, mRNA and translations from feature tables.
compseq	Counts the composition of dimer/trimer/etc words in a sequence.
cons	Creates a consensus from multiple alignments.
contacts	Reads coordinate files and writes files of intra-chain residue-residue contact data.
cpgplot	Plot CpG rich areas.
cpgreport	Reports all CpG rich regions.
cusp	Create a codon usage table.
cutgextract	Extract data from CUTG.
cutseq	Removes a specified section from a sequence.
dan	Calculates DNA RNA/DNA melting temperature.

Table 12-8. EMBOSS program descriptions (continued)

Program	Description
dbiblast	Index a BLAST database.
dbifasta	Index a FASTA database.
dbiflat	Index a flat file database.
dbigcg	Index a GCG formatted database.
degapseq	Removes gap characters from sequences.
descseq	Alter the name or description of a sequence.
dichet	Parse dictionary of heterogen groups.
diffseq	Find differences (SNPs) between nearly identical sequences.
digest	Protein proteolytic enzyme or reagent cleavage digest.
distmat	Creates a distance matrix from multiple alignments.
domainer	Reads protein coordinate files and writes domains coordinate files.
dotmatcher	Displays a thresholded dotplot of two sequences.
dotpath	Displays a non-overlapping wordmatch dotplot of two sequences.
dottup	Displays a wordmatch dotplot of two sequences.
dreg	Regular expression search of a nucleotide sequence.
einverted	Finds DNA inverted repeats.
embossdata	Finds or fetches the data files read in by the EMBOSS programs.
embossversion	Writes the current EMBOSS version number.
emma	Multiple alignment program—interface to ClustalW program.
emowse	Protein identification by mass spectrometry.
entret	Reads and writes (returns) flat file entries.
eprimer3	Picks PCR primers and hybridization oligos.
equicktandem	Finds tandem repeats.
est2genome	Align EST and genomic DNA sequences.
etandem	Looks for tandem repeats in a nucleotide sequence.
extractfeat	Extract features from a sequence.
extractseq	Extract regions from a sequence.
findkm	Find Km and Vmax for an enzyme reaction by a Hanes/Woolf plot.
freak	Residue/base frequency table or plot.
funky	Reads clean coordinate files and writes file of protein-heterogen contact data.
fuzznuc	Nucleic acid pattern search.
fuzzpro	Protein pattern search.
fuzztran	Protein pattern search after translation.
garnier	Predicts protein secondary structure.
geecee	Calculates the fractional GC content of nucleic acid sequences.
getorf	Finds and extracts open reading frames (ORFs).
groups	Removes redundant hits from a scop families file.
helixturnhelix	Report nucleic acid binding motifs.
hetparse	Converts raw dictionary of heterogen groups to a file in EMBL-like format.
hmmgen	Generates a hidden Markov model for each alignment in a directory.
hmoment	Hydrophobic moment calculation.

Table 12-8. EMBOSS program descriptions (continued)

Program	Description
iep	Calculates the isoelectric point of a protein.
infoalign	Information on a multiple sequence alignment.
infoseq	Displays some simple information about sequences.
interface	Reads coordinate files and writes files of inter-chain residue-residue contact data.
isochore	Plots isochores in large DNA sequences.
lindna	Draws linear maps of DNA constructs.
listor	Writes a list file of the logical OR of two sets of sequences.
marscan	Finds MAR/SAR sites in nucleic sequences.
maskfeat	Mask off features of a sequence.
maskseq	Mask off regions of a sequence.
matcher	Finds the best local alignments between two sequences.
megamerger	Merge two large overlapping nucleic acid sequences.
merger	Merge two overlapping nucleic acid sequences.
msbar	Mutate sequence beyond all recognition.
mwcontam	Shows molecular weights that match across a set of files.
mwfilter	Filter noisy molwts from mass spec output.
needle	Needleman-Wunsch global alignment.
newcpgreport	Report CpG rich areas.
newcpgseek	Reports CpG rich regions.
newseq	Type in a short new sequence.
noreturn	Removes carriage return from ASCII files.
notseq	Excludes a set of sequences and writes out the remaining ones.
nrscope	Converts redundant EMBL-format SCOP file to non-redundant one.
nthseq	Writes one sequence from a multiple set of sequences.
octanol	Displays protein hydropathy.
oddcomp	Finds protein sequence regions with a biased composition.
palindrome	Looks for inverted repeats in a nucleotide sequence.
pasteseq	Insert one sequence into another.
patmatdb	Search a protein sequence with a motif.
patmatmotifs	Search a PROSITE motif database with a protein sequence.
pdbparse	Parses PDB files and writes cleaned-up protein coordinate files.
pdbtosp	Convert raw SWISS-PROT:PDB equivalence file to EMBL-like format.
pepcoil	Predicts coiled coil regions.
pepinfo	Plots simple amino acid properties in parallel.
pepnet	Displays proteins as a helical net.
pepstats	Protein statistics.
pepwheel	Shows protein sequences as helixes.
pepwindow	Displays protein hydropathy.
pepwindowall	Displays protein hydropathy of a set of sequences.
plotcon	Plots the quality of conservation of a sequence alignment.
plotorf	Plot potential open reading frames.

Table 12-8. EMBOSS program descriptions (continued)

Program	Description
polydot	Displays all-against-all dotplots of a set of sequences.
preg	Regular expression search of a protein sequence.
prettyplot	Displays aligned sequences, with coloring and boxing.
prettyseq	Output sequence with translated ranges.
primersearch	Searches DNA sequences for matches with primer pairs.
printsextract	Extract data from PRINTS.
profgen	Generates various profiles for each alignment in a directory.
profit	Scan a sequence or database with a matrix or profile.
prophecy	Creates matrices/profiles from multiple alignments.
prophet	Gapped alignment for profiles.
prosextract	Builds the PROSITE motif database for *patmatmotifs* to search.
pscan	Scans proteins using PRINTS.
psiblasts	Runs PSI-BLAST given *scopalign* alignments.
rebaseextract	Extract data from REBASE.
recoder	Remove restriction sites but maintain the same translation.
redata	Search REBASE for enzyme name, references, suppliers etc.
remap	Display a sequence with restriction cut sites, translation etc.
restover	Finds restriction enzymes that produce a specific overhang.
restrict	Finds restriction enzyme cleavage sites.
revseq	Reverse and complement a sequence.
scopalign	Generate alignments for families in a scop classification file by using STAMP.
scope	Convert raw scop classification file to EMBL-like format.
scopnr	Removes redundant domains from a scop classification file.
scopparse	Converts raw scop classification files to a file in EMBL-like format.
scoprep	Reorder scop classificaiton file so that the representative structure of each family is given first.
scopreso	Removes low resolution domains from a scop classification file.
scopseqs	Adds PDB and SWISS-PROT sequence records to a scop classification file.
seealso	Finds programs sharing group names.
seqalign	Generate extended alignments for families in a scop families file by using ClustalW with seed alignments.
seqmatchall	Does an all-against-all comparison of a set of sequences.
seqret	Reads and writes (returns) sequences.
seqretsplit	Reads and writes (returns) sequences in individual files.
seqsearch	Generate files of hits for families in a scop classification file by using PSI-BLAST with seed alignments.
seqsort	Reads multiple files of hits and writes a non-ambiguous file of hits (scop families file) plus a validation file.
seqwords	Generate file of hits for scop families by searching SWISS-PROT with keywords.
showalign	Displays a multiple sequence alignment.
showdb	Displays information on the currently available databases.
showfeat	Show features of a sequence.
showorf	Pretty output of DNA translations.

Table 12-8. EMBOSS program descriptions (continued)

Program	Description
showseq	Display a sequence with features, translation, etc.
shuffleseq	Shuffles a set of sequences maintaining composition.
sigcleave	Reports protein signal cleavage sites.
siggen	Generates a sparse protein signature from an alignment and residue contact data.
sigscan	Scans a signature against SWISS-PROT and writes a signature hits files.
silent	Silent mutation restriction enzyme scan.
skipseq	Reads and writes (returns) sequences, skipping the first few.
splitter	Split a sequence into (overlapping) smaller sequences.
stretcher	Finds the best global alignment between two sequences.
stssearch	Searches a DNA database for matches with a set of STS primers.
supermatcher	Finds a match of a large sequence against one or more sequences.
swissparse	Retrieves sequences from SWISS-PROT using keyword search.
syco	Synonymous codon usage Gribskov statistic plot.
textsearch	Search sequence documentation text; SRS and Entrez are faster.
tfextract	Extract data from TRANSFAC.
tfm	Displays a program's help documentation manual.
tfscan	Scans DNA sequences for transcription factors.
tmap	Displays membrane spanning regions.
tranalign	Align nucleic coding regions given the aligned proteins.
transeq	Translate nucleic acid sequences.
trimest	Trim poly-A tails off EST sequences.
trimseq	Trim ambiguous bits off the ends of sequences.
union	Reads sequence fragments and builds one sequence.
vectorstrip	Strips out DNA between a pair of vector sequences.
water	Smith-Waterman local alignment.
whichdb	Search all databases for an entry.
wobble	Wobble base plot.
wordcount	Counts words of a specified size in a DNA sequence.
wordmatch	Finds all exact matches of a given size between 2 sequences.
wossname	Finds programs by keywords in their one-line documentation.
yank	Reads a sequence range, appends the full USA to a list file.

Details of EMBOSS Programs

The programs are listed alphabetically. A description, example, and summary of the command-line arguments are given. Optional and advanced qualifiers are listed if applicable.

General qualifiers: (used by all programs)

-help (boolean)
: Report command line options. More information on associated and general qualifiers can be found with `-help -verbose`.

aaindexextract

aaextractindex extracts information from the AAINDEX database for use by *pepwindow* and *pepwindowall*. AAINDEX is a database containing properties of amino acids. The AAINDEX database file *aaindex1* can be downloaded from *ftp://ftp.genome.ad.jp/pub/db/genomenet/aaindex/aaindex1*.

Here is a sample session with *aaindexextract*:

```
% aaindexextract
Extract data from AAINDEX
Full pathname of file aaindex1: ~/aaindex1
```

Mandatory qualifiers:

[-inf] (infile)
    Full pathname of file *aaindex1*.

EMBOSS

abiview

abiview reads an ABI sequence trace file and displays the results as a graphic.

Here is a sample session with *abiview*:

```
% abiview
Name of the ABI trace file: ba16d2.s1
Output sequence [outfile.fasta]:
Graph type [x11]:
```

Mandatory qualifiers:

[-fname] (infile)
    Name of the ABI trace file.

[-outseq] (seqout)
    Sequence file.

-graph (xygraph)
    Graph type.

Optional qualifiers:

-startbase (integer)
    First base to report or display.

-endbase (integer)
    Last sequence base to report or display. If the default is set to zero, the value of this qualifier is taken as the maximum number of bases.

-yticks (boolean)
    Display y-axis ticks.

-[no]sequence (boolean)
    Display the sequence on the graph.

-window (integer)
    Sequence display window size.

-bases (string)
    Base graphs to be displayed.

Advanced qualifiers:

-separate (boolean)
: Separate the trace graphs for the 4 bases.

alignwrap

alignwrap aligns a set of sequences to a seed alignment. This package is still being developed.

Here is a sample session with *alignwrap*:

```
% alignwrap
```

Mandatory qualifiers:

-inpath (string)
: Directory containing the seed alignments (input).

-extn (string)
: File extension of seed alignment files (input).

-scopfamilies (string)
: scop families file containing the set of sequences in EMBL-like format.

-outpath (string)
: Directory for extended alignments (output).

-outextn (string)
: File extension of extended alignment files (output).

antigenic

antigenic predicts potentially antigenic regions of a protein sequence.

Here is a sample session with *antigenic*:

```
% antigenic
Finds antigenic sites in proteins
Input sequence: sw:act1_fugru
Minimum length [6]:
Output file [act1_fugru.antigenic]:
```

Mandatory qualifiers:

[-sequence] (seqall)
: Sequence database USA.

-minlen (integer)
: Minimum length.

[-outfile] (report)
: Output report filename.

backtranseq

backtranseq takes a protein sequence and makes a best estimate of the likely nucleic acid sequence it may have come from.

Here is a sample session with *backtranseq*. Note that this is a human protein, so the default (human) codon frequency file is used:

```
% backtranseq
Back translate a protein sequence
Input sequence: sw:opsd_human
Output sequence [opsd_human.fasta]:
```

Here is a session using a drosophila sequence and codon table:

```
% backtranseq -cfile Edrosophila.cut
Back translate a protein sequence
Input sequence: sw:ach2_drome
Output sequence [ach2_drome.fasta]:
```

Mandatory qualifiers:

[-sequence] (sequence)
    Sequence USA.

[-outfile] (seqout)
    Output sequence USA.

Optional qualifiers:

-cfile (codon)
    Codon usage table name.

EMBOSS

banana

banana predicts bending of a normal (B) DNA double helix.

Here is a sample session with *banana*:

```
% banana embl:rnu68037
Bending and curvature plot in B-DNA
Graph type [x11]:
```

Mandatory qualifiers (bold if not always prompted):

[-sequence] (sequence)
    Sequence USA.

-graph *(graph)*
    Graph type.

Optional qualifiers:

-anglesfile (datafile)
    Angles file.

-residuesperline (integer)
    Number of residues to display on each line.

-outfile (outfile)
    Output filename.

Advanced qualifiers:

-data (boolean)
    Output as data.

biosed

biosed is a simple sequence editing utility that searches for a target subsequence in one or more input sequences and replaces it with a specified second subsequence (or optionally just deletes the found target subsequence). *biosed* was inspired by the useful Unix utility *sed*.

Here is a sample session with *biosed* that replaces all "T's" with "U's" to create an RNA sequence:

```
% biosed em:hsfau hsfau.rna -target T -replace U
```

Replace all "RGD" protein motifs with "XXRGDXX":

```
% biosed sw:A4M1_HUMAN A4M1_HUMAN.pep -target RGD -replace XXRGDXX
```

Mandatory qualifiers (bold if not always prompted):

[-sequence] (seqall)
  Sequence database USA.

-target (string)
  Sequence section to match.

-replace *(string)*
  Replacement sequence section.

[-outseq] (seqout)
  Output sequence USA.

Advanced qualifiers:

-delete (boolean)
  Delete the target sequence sections.

btwisted

btwisted takes a region of a pure DNA sequence and calculates by simple arithmetic the probable overall twist of the sequence and the stacking energy.

Here is a sample session with *btwisted*:

```
% btwisted embl:ab000095 -sbegin 100 -send 120
Calculates the twisting in a B-DNA sequence
Output file [ab000095.btwisted]:
```

Mandatory qualifiers:

[-sequence] (sequence)
  Sequence USA.

[-outfile] (outfile)
  Output filename.

Advanced qualifiers:

-angledata (string)
  File containing base pair twist angles.

-energydata (string)
  File containing base pair stacking energies.

cai

cai calculates the Codon Adaptation Index. It is a simple, effective measure of synonymous codon usage bias.

Here is a sample session with *cai*:

```
% cai embl:AB009602
CAI codon adaptation index
Codon usage file [Eyeastcai.cut]:
Output file [ab009602.cai]:
```

Mandatory qualifiers:

[-seqall] (seqall)
: Sequence database USA.

-cfile (codon)
: Codon usage file.

[-outfile] (outfile)
: Output filename.

chaos

chaos creates a game representation plot for a sequence.

Here is a sample session with *chaos*:

```
% chaos embl:eclac
Create a chaos game representation plot for a sequence
Graph type [x11]:
```

Mandatory qualifiers (bold if not always prompted):

[-sequence] (sequence)
: Sequence USA.

-graph *(graph)*
: Graph type.

-outfile *(outfile)*
: Output filename.

Advanced qualifiers:

-data (boolean)
: Display as data.

charge

charge reads a protein sequence and writes a file (or plots a graph) of the charges to the amino acids within a window of specified length as the window is moved along the sequence.

Here is a sample session with *charge*:

```
% charge sw:hbb_human
Protein charge plot
Output file [hbb_human.charge]:
```

Mandatory qualifiers (bold if not always prompted):

[-seqall] (seqall)
: Sequence database USA.

-graph *(xygraph)*
: Graph type.

-outfile *(outfile)*
: Output filename.

Optional qualifiers:

-window (integer)
: Window.

Advanced qualifiers:

-aadata (string)
: Amino acid property data filename.

-plot (boolean)
: Produce graphic.

---

checktrans

checktrans reads a protein sequence containing stops and writes a report of any open reading frames greater than a minimum size. The input sequence is typically produced by *transeq*.

Here is a sample session with *checktrans*, using the output from a *transeq* run:

```
% transeq embl:paamir paamir.pep -auto
% checktrans
Input sequence: paamir.pep
Minimum ORF Length to report [100]: 30
Output file [paamir_1.checktrans]:
Output sequence [paamir_1.fasta]:
```

Mandatory qualifiers:

[-sequence] (seqall)
: Sequence database USA.

-orfml (integer)
: Minimum ORF Length to report.

[-report] (outfile)
: Output filename.

-outseq (seqoutall)
: Sequence file to hold output ORF sequences.

Advanced qualifiers:

-featout (featout)
: File for output features.

chips

chips calculates Frank Wright's Nc statistic for the effective number of codons used.

Here is a sample session with *chips*. If the sequence extends beyond the coding region, the start and/or end positions of the CDS must be provided. *chips* analyzes only protein coding regions:

```
% chips -sbeg 135 -send 1292
Input sequence: embl:paamir
Output file [paamir.chips]:
```

Mandatory qualifiers:

[-seqall] (seqall)
  Sequence database USA.

[-outfile] (outfile)
  Output filename.

Advanced qualifiers:

-cfile (codon)
  Codon usage file.

cirdna

cirdna draws circular maps of DNA constructs.

Here is a sample session with *cirdna*:

```
% cirdna
Draws circular maps of DNA constructs
Graph type [x11]:
Input file [inputfile]: data.cirp
do you want a ruler (Y or N) [Y]:
type of blocks (enter Open, Filled, or Outline) [Filled]:
ticks inside or outside the circle (enter In or Out) [Out]:
text inside or outside the blocks (enter In or Out) [In]:
```

Mandatory qualifiers:

-graphout (graph)
  Graph type.

-inputfile (infile)
  Input file containing mapping data.

-ruler (string)
  Do you want a ruler (`Y` or `N`)?

-blocktype (string)
  Type of blocks: `Open`, `Gilled`, or `Outline`. `Outline` draws filled blocks surrounded by a black border.

-posticks (string)
  Ticks inside or outside the circle (enter `In` or `Out`).

-posblocks (string)
  Text inside or outside the blocks (enter `In` or `out`).

Optional qualifiers:

-originangle (float)
: Position of the molecule's origin on the circle (enter a number in the range 0–360).

-intersymbol (string)
: Do you want horizontal junctions between blocks (Y or N)?

-intercolor (integer)
: Color of junctions between blocks (enter a color number).

-interticks (string)
: Do you want horizontal junctions between ticks (Y or N)?

-gapsize (integer)
: Interval between ticks in the ruler (enter an integer).

-ticklines (string)
: Do you want vertical lines at the ruler's ticks (Y or N)?

-textheight (float)
: Height of text; enter a number less than or greater than one to decrease or increase the size, respectively.

-textlength (float)
: Length of text; enter a number less than or greater than one to decrease or increase the size, respectively.

-tickheight (float)
: Height of ticks; enter a number less than or greater than one to decrease or increase the size, respectively.

-blockheight (float)
: Height of blocks; enter a number less than or greater than one to decrease or increase the size, respectively.

-rangeheight (float)
: Height of range ends; enter a number less than or greater than one to decrease or increase the size, respectively.

-gapgroup (float)
: Space between groups; enter a number less than or greater than one to decrease or increase the size, respectively.

-postext (float)
: Space between text and ticks, blocks, and ranges; enter a number less than or greater than one to decrease or increase the size, respectively.

codcmp

codcmp reads and compares two codon usage table files.

Here is a sample session with *codcmp*, comparing the codon usage tables for *Escherichia coli* and *Haemophilus influenzae*:

```
% codcmp
Codon usage file [Ehum.cut]: Eeco.cut
Codon usage file [Ehum.cut]: Ehin.cut
Output file [outfile.codcmp]:
```

Mandatory qualifiers:

[-first] (codon)
: First codon usage file.

[-second] (codon)
: Second codon usage file.

[-outfile] (outfile)
: Output filename.

---

coderet

coderet extracts CDS, mRNA and translations from feature tables.

Here is a sample session with *coderet*. To extract all of the CDS, mRNA and the protein translations, type:

```
% coderet
Extract CDS, mRNA and translations from feature tables
Input sequence(s): embl:X03487
Output sequence [hsferg1.fasta]:
```

Mandatory qualifiers:

[-seqall] (seqall)
: Sequence database USA.

[-seqout] (seqout)
: Output sequence USA.

Advanced qualifiers:

-[no]cds (boolean)
: Extract CDS sequences.

-[no]mrna (boolean)
: Extract mrna sequences.

-[no]translation (boolean)
: Extract translated sequences.

---

compseq

compseq counts the composition of dimer/trimer/etc words in a sequence.

Here is a sample session with *compseq*. To count the frequencies of dinucleotides in a file:

```
% compseq embl:hsfau 2 result3.comp
```

To count the frequencies of hexanucleotides, without outputting the results of hexanucleotides that do not occur in the sequence:

```
% compseq embl:hsfau 6 result6.comp -nozero
```

To count the frequencies of trinucleotides in frame 2 of a sequence using a previously prepared *compseq* output to show the expected frequencies:

```
% compseq embl:hsfau 3 result3.comp -frame 2 -in prev.comp
```

EMBOSS

Mandatory qualifiers:

[-sequence] (seqall)
: Sequence database USA.

[-word] (integer)
: The size of word (n-mer) to count. If you want to count codon frequencies, enter `3` here.

[-outfile] (outfile)
: The results file.

Optional qualifiers (bold if not always prompted):

-infile (infile)
: This is a file previously produced by *compseq* that can be used to set the expected frequencies of words in an analysis. The word size in the current run must be the same as the word size in this results file. Obviously, you should use a file produced from protein sequences if you are counting protein sequence word frequencies, or a file made from nucleotide frequencies if you are analyzing a nucleotide sequence.

-frame (integer)
: The normal behavior of *compseq* is to count the frequencies of all words that occur by moving a window of length `word` up by one each time. This option allows you to move the window up by the length of the word each time, skipping intervening words. You can count only those words that occur in a single frame of the word by setting this value to a number other than `0`. If you set it to `1` it will only count the words in frame 1, `2` will only count the words in frame 2 and so on.

-**[no]ignorebz** *(boolean)*
: The amino acid code B represents Asparagine or Aspartic acid, and the code Z represents Glutamine or Glutamic acid. These codes are not commonly used, and you may not want to count words containing them. This command will note codes B and Z in the count of "Other" words.

-**reverse** *(boolean)*
: Set this option to true if you want to count words in the reverse complement of a nucleic sequence.

-[no]zerocount (boolean)
: You can make the output results file much smaller if you do not display the words with a zero count.

---

cons

cons calculates a consensus sequence from a multiple sequence alignment.

Here is a sample session with *cons*:

```
% cons
Creates a consensus from multiple alignments
Input sequence set: aligned.fasta
Output file [outfile.cons]: aligned.cons
```

Mandatory qualifiers:

[-msf] (seqset)
: File containing a sequence alignment.

[-outseq] (seqout)
: Output sequence USA.

Optional qualifiers:

-datafile (matrix)
: This is the scoring matrix file used when comparing sequences. By default, it is the file *EBLOSUM62* (for proteins) or the file *EDNAFULL* (for nucleic sequences). These files are found in the *data* directory of the EMBOSS installation.

-plurality (float)
: Set a cutoff for the number of positive matches below which there is no consensus. The default plurality is taken as half the total weight of all the sequences in the alignment.

-identity (integer)
: Provides the facility of setting the required number of identities at a site for it to give a consensus at that position. If this qualifier is set to the number of sequences in the alignment, only columns of identities contribute to the consensus.

-name (string)
: Name of the consensus sequence.

-setcase (float)
: Sets the threshold for the positive matches. Residues in the consensus sequence have their case determined by this threshold: those above it are in uppercase; those below it are in lowercase.

---

contacts

contacts reads coordinate files and writes files of intra-chain residue-residue contact data.

Here is a sample session with *contacts*:

```
% contacts
Reads coordinate files and writes contact files
Location of coordinate files for input (embl-like format) [./]:
Extension of coordinate files (embl-like format) [.pxyz]:
Location of contact files for output [./]:
Extension of contact files [.con]:
Threshold contact distance [1.0]:
Name of data file with van der Waals radii [Evdw.dat]:
Name of log file for the build [contacts.log]:
```

Mandatory qualifiers:

[-cpdb] (string)
: Location of coordinate files (EMBL format input).

[-cpdbextn] (string)
: Extension of coordinate files.

-vdwf (string)
: Name of data file with Van der Waals radii.

EMBOSS

-thresh (float)
Threshold contact distance.

[-con] (string)
Location of contact files for output.

[-conextn] (string)
Extension of contact files.

-conerrf (outfile)
Name of log file for the build.

Advanced qualifiers:

-ignore (float)
If any two atoms from two different residues are at least this distance apart, no further interatomic contacts are checked for that residue pair. This option speeds the calculation up considerably.

cpgplot

cpgplot plots CpG rich areas.

Here is a sample session with *cpgplot*:

```
% cpgplot embl:rnu68037
Plot CpG rich areas
Window size [100]:
Shift increment [1]:
Minimum Length [200]:
Minimum observed/expected [0.6]:
Minimum percentage [50.]:
Output file [rnu68037.cpgplot]:
Graph type [x11]:
```

Mandatory qualifiers:

[-sequence] (seqall)
Sequence database USA.

-window (integer)
The percentage CG content and the observed frequency of CG is calculated within a window whose size is set by this parameter. The window is moved down the sequence and these statistics are calculated at each position the window is moved to.

-shift (integer)
This determines the number of bases the window is moved each time after values of the percentage CG content and the observed frequency of CG are calculated within the window.

-minlen (integer)
This sets the minimum length that a CpG island must be before it is reported.

-minoe (float)
This sets the minimum average observed to expected ratio of C plus G to CpG in a set of 10 windows that are required before a CpG island is reported.

-minpc (float)
: This sets the minimum average percentage of G plus C: a set of 10 windows required before a CpG island is reported.

[-outfile] (outfile)
: This sets the name of the file holding the report of the input sequence name, CpG island parameters, and the output details of any CpG islands that are found.

[-graph] (xygraph)
: Graph type.

Advanced qualifiers:

-[no]obsexp (boolean)
: If set to true, the graph of the observed to expected ratio of C plus G to CpG within a window is displayed.

-[no]cg (boolean)
: If set to true, the graph of the regions determined to be CpG islands is displayed.

-[no]pc (boolean)
: If set to true, the graph of the percentage C plus G within a window is displayed.

-featout (featout)
: File for output features.

---

cpgreport

cpgreport scans a nucleotide sequence for regions with higher than expected frequencies of the dinucleotide CG.

Here is a sample session with *cpgreport*:

```
% cpgreport embl:rnu68037
Reports CpG rich regions
CpG score [17]:
Output file [rnu68037.cpgreport]:
```

Mandatory qualifiers:

[-sequence] (seqall)
: Sequence database USA.

-score (integer)
: This sets the score for each CG sequence found. A value of `17` is more sensitive, but `28` has also been used with some success.

[-outfile] (outfile)
: Output filename.

Advanced qualifiers:

-featout (featout)
: File for output features.

---

cusp

cusp reads one or more coding sequences (CDS sequence only) and calculates a codon frequency table. The output file can be used as a codon usage table in other applications.

Here is a sample session with *cusp*, using just one sequence. For normal use, extract a set of coding sequences and use them as input:

```
% cusp -sbeg 135 -send 1292
Create a codon usage table
Input sequence: embl:paamir
Output file [paamir.cusp]:
```

Mandatory qualifiers:

[-sequence] (seqall)
Sequence database USA.

[-outfile] (outfile)
Output filename.

Advanced qualifiers:

-cfile (codon)
Codon usage table name.

cutgextract

cutgextract extracts data from CUTG.

Here is a sample session with *cutgextract*:

```
% cutgextract
```

Mandatory qualifiers:

[-directory] (dirlist)
CUTG directory.

Advanced qualifiers:

-wildspec (string)
Type of codon file.

cutseq

cutseq allows you to cut out a region from your sequence by specifying the begin and end positions of the sequence to remove.

To remove bases 10 to 12 from a sequence *gatta.seq* and write to the new sequence file *gatta2.seq*:

```
% cutseq gatta.seq gatta2.seq -from=10 -to=12
```

To remove the first 20 bases from *hatta.seq* and write it to *jsh.seq*:

```
% cutseq -seq=hatta.seq -from=1 -to=20 -out=jsh.seq
```

Mandatory qualifiers:

[-sequence] (sequence)
Sequence USA.

-from (integer)
This is the start position (inclusive) of the section of the sequence that you want to remove.

-to (integer)
: This is the end position (inclusive) of the section of the sequence that you want to remove.

[-outseq] (seqout)
: Output sequence USA.

dan

dan calculates the melting temperature (Tm) and the percent G+C of a nucleic acid sequence (optionally plotting them).

Here is a sample session with *dan*:

```
% dan
Input sequence: embl:paamir
Enter window size [20]:
Enter Shift Increment [1]:
Enter DNA concentration (nM) [50.]:
Enter salt concentration (mM) [50.]:
Output file [paamir.dan]:
```

An example of producing a plot of Tm:

```
% dan -plot
Input sequence(s): embl:paamir
Enter window size [20]:
Enter Shift Increment [1]:
Enter DNA concentration (nM) [50.]:
Enter salt concentration (mM) [50.]:
Enter minimum temperature [55.]:
Graph type [x11]:
```

Mandatory qualifiers (bold if not always prompted):

[-sequence] (seqall)
: Sequence database USA.

-windowsize (integer)
: The values of melting point and other thermodynamic properties of the sequence are determined by taking a short length of sequence known as a window and determining the properties of the sequence in that window. The window is incrementally moved along the sequence, and the properties are calculated at each new position.

-shiftincrement (integer)
: This is the amount the window is moved at each increment in order to find the melting point and other properties along the sequence.

-dnaconc (float)
: Enter DNA concentration (nM).

-saltconc (float)
: Enter salt concentration (mM).

-formamide *(float)*
: This specifies the percent formamide to be used in calculations (ignored unless `-product` is used).

-mismatch *(float)*
: This specifies the percent mismatch to be used in calculations (ignored unless `-product` is used).

-prodlen *(integer)*
: This specifies the product length to be used in calculations (ignored unless `-product` is used).

-mintemp *(float)*
: Enter a minimum value for the temperature scale (y-axis) of the plot.

-graph *(xygraph)*
: Graph type.

-outfile *(report)*
: If a plot is not being produced, data (on the melting point etc., in each window along the sequence) is output to this file.

Optional qualifiers (bold if not always prompted):

-temperature *(float)*
: If `-thermo` has been specified, this specifies the temperature at which to calculate the DeltaG, DeltaH and DeltaS values.

Advanced qualifiers:

-rna (boolean)
: This specifies that the sequence is an RNA sequence, not a DNA sequence.

-product (boolean)
: This prompts for percent formamide, percent of mismatches allowed, and product length.

-thermo (boolean)
: Output the DeltaG, DeltaH, and DeltaS values of the sequence windows to the output data file.

-plot (boolean)
: If this is not specified, the file of output data is produced, else a plot of the melting point along the sequence is produced.

---

dbiblast

dbiblast indexes a BLAST database created by the NCBI indexing programs (*formatdb*, *pressdb*, or *setdb*) and builds EMBL CD-ROM format index files.

Here is a sample session with *dbiblast*, using a BLAST2 protein database:

```
% dbiblast
      NCBI : NCBI with | characters
       GCG : GCG format dbname:entryname
    SIMPLE : ID and accession
        ID : entryname
   UNKNOWN : unknown
Entry format [unknown]: gcg
Database name: swnew
Database directory [.]: /nfs/disk42/pmr/emboss/test/blastp2/
database base filename [swnew]:
Release number [0.0]:
Index date [00/00/00]: 04/02/00
```

```
          N : nucleic
          P : protein
          ? : unknown
Sequence type [unknown]: p
          1 : wublast and setdb/pressdb
          2 : formatdb
          0 : unknown
Blast index version [unknown]: 2
```

Mandatory qualifiers:

[-dbname] (string)
Database name.

-directory (string)
Database directory.

-filenames (string)
Wildcard database filename.

-release (string)
Release number.

-date (string)
Index date.

-seqtype (menu)
Sequence type.

-blastversion (menu)
BLAST index version.

Advanced qualifiers:

-fields (menu)
Index fields.

-exclude (string)
Wildcard filename(s) to exclude.

-indexdirectory (string)
Index directory.

-sortoptions (string)
Sort options, typically `-T .` to use current directory for work files, or `-k 1,1` to force GNU sort to use the first field.

-maxindex (integer)
Maximum index length.

-[no]systemsort (boolean)
Use system sort utility.

-[no]cleanup (boolean)
Clean up temporary files.

-sourcefile (boolean)
Use FASTA source file.

dbifasta

dbifasta indexes a FASTA format flat file database of one or more files, and builds EMBL CD-ROM format index files.

Here is a sample session with *dbifasta*, using data in the *test/wormpep* directory of the distribution (normally indexed as *wormpep*):

```
% dbifasta
Index a fasta database
    simple : >ID
     idacc : >ID ACC
     gcgid : >db:ID
  gcgidacc : >db:ID ACC
      ncbi : >blah|...[|ACC]|ID
ID line format [idacc]: idacc
Database name: wormpep
Database directory [.]: /data/worm/
Wildcard database filename [*.dat]: wormpep
Release number [0.0]: 1.0
Index date [00/00/00]: 04/02/00
```

Mandatory qualifiers:

-idformat (menu)
  ID line format.

-directory (string)
  Database directory.

-filenames (string)
  Wildcard database filename.

[-dbname] (string)
  Database name.

-release (string)
  Release number.

-date (string)
  Index date.

Advanced qualifiers:

-fields (menu)
  Index fields.

-exclude (string)
  Wildcard filename(s) to exclude.

-indexdirectory (string)
  Index directory.

-maxindex (integer)
  Maximum index length.

-sortoptions (string)
  Sort options, typically `-T .` to use current directory for work files, or `-k 1,1` to force GNU sort to use the first field.

-[no]systemsort (boolean)
Use system sort utility.

-[no]cleanup (boolean)
Clean up temporary files.

dbiflat

dbiflat indexes a flat file database of one or more files and builds EMBL CD-ROM format index files.

Here is a sample session with *dbiflat*, using the data in the *test/embl* directory of the distribution (normally indexed as *tembl*):

```
% dbiflat
       EMBL : EMBL
      SWISS : Swiss-Prot, SpTrEMBL, TrEMBLnew
         GB : Genbank, DDBJ
Entry format [SWISS]: EMBL
Database name: tembl
Database directory [.]: /nfs/disk42/pmr/emboss/osf/emboss/test/embl/
Wildcard database filename [*.dat]:
Release number [0.0]: 1.0
Index date [00/00/00]: 04/02/00
```

Mandatory qualifiers:

-idformat (menu)
Entry format.

-directory (string)
Database directory.

-filenames (string)
Wildcard database filename.

[-dbname] (string)
Database name.

-release (string)
Release number.

-date (string)
Index date.

Advanced qualifiers:

-fields (menu)
Index fields.

-exclude (string)
Wildcard filename(s) to exclude.

-indexdirectory (string)
Index directory.

-maxindex (integer)
Maximum index length.

-sortoptions (string)
Sort options, typically `-T .` to use current directory for work files, or `-k 1,1` to force GNU sort to use the first field.

-[no]systemsort (boolean)
  Use system sort utility.

-[no]cleanup (boolean)
  Clean up temporary files.

dbigcg

dbigcg indexes a GCG-format database of one or more files and builds EMBL CD-ROM format index files.

Here is a sample session with *dbigcg*:

```
% dbigcg
         EMBL : EMBL
        SWISS : Swiss-Prot, SpTrEMBL, TrEMBLnew
           GB : Genbank, DDBJ
          PIR : NBRF
Entry format [EMBL]:
Database name: tembl
Database directory [.]: /nfs/disk42/pmr/emboss/test/gcgembl/
Wildcard database filename [*.seq]:
Release number [0.0]:
Index date [00/00/00]: 04/02/00
```

Mandatory qualifiers:

-idformat (menu)
  Entry format.

-directory (string)
  Database directory.

-filenames (string)
  Wildcard database filename.

[-dbname] (string)
  Database name.

-release (string)
  Release number.

-date (string)
  Index date.

Advanced qualifiers:

-fields (menu)
  Index fields.

-exclude (string)
  Wildcard filename(s) to exclude.

-indexdirectory (string)
  Index directory.

-maxindex (integer)
  Maximum index length.

sortoptions (string)
  Sort options, typically `-T .` to use current directory for work files, or `-k 1,1` to force GNU sort to use the first field.

-[no]systemsort (boolean)
Use system sort utility.

-[no]cleanup (boolean)
Clean up temporary files.

---

degapseq

degapseq reads one or more sequences, eliminates gap characters, and writes them out again. This process removes gaps from aligned sequences.

Here is a sample session with *degapseq*:

```
% degapseq alignment.seq nogaps.seq
```

Mandatory qualifiers:

[-sequence] (seqall)
Sequence database USA.

[-outseq] (seqoutall)
Output sequence(s) USA.

EMBOSS

---

descseq

descseq allows you to change the name or description of a sequence, then write out a new file containing the changes. The original sequence is unaltered.

Set the name of a sequence to "myclone23":

```
% descseq -seq clone23 -out clone23.seq -name "myclone23"
```

Set the description of a sequence to "This is my clone number 244":

```
% descseq -seq xy24 -out xy24.seq -desc "This is my clone number 244"
```

Append some text to the description of a sequence:

```
% descseq -seq est.seq -out est4.seq -desc " (submitted)" -append
```

Mandatory qualifiers:

[-sequence] (sequence)
Sequence USA.

[-outseq] (seqout)
Output sequence USA.

Optional qualifiers:

-name (string)
Name of the sequence.

-description (string)
Description of the sequence.

Advanced qualifiers:

-append (boolean)
This allows you to append a name or description to the existing name or description of a sequence.

dichet

dichet is a parse dictionary of heterogen groups.

Here is a sample session with *dichet*:

```
% dichet
```

Mandatory qualifiers (bold if not always prompted):

[-infl] (infile)
: Name of raw input file (dictionary of heterogen groups).

[-dogrep] (boolean)
: Search a directory of files with keywords.

[-outfl] (outfile)
: Name of output file.

-path *(string)*
: Directory to search with keywords.

-extn *(string)*
: Exension of files to search in above directory.

diffseq

diffseq takes two overlapping, nearly identical sequences and reports the differences between them, together with any features that overlap with these regions.

Here is a sample session with *diffseq*:

```
% diffseq tembl:ap000504 tembl:af129756
Find differences (SNPs) between nearly identical sequences
Word size [10]:
Output file [ap000504.diffseq]:
```

Mandatory qualifiers:

[-asequence] (sequence)
: Sequence USA.

[-bsequence] (sequence)
: Sequence USA.

-wordsize (integer)
: Word size.

-outfile (report)
: Output report file.

Optional qualifiers:

-afeatout (featout)
: File for output of first sequence's normal tab delimited gff's.

-bfeatout (featout)
: File for output of second sequence's normal tab delimited gff's.

-columns (boolean)
: The default format for the output report file is to have several lines per difference giving the sequence positions, sequences, and features. If this option is set to true, the output report file format is changed to a set of columns and no feature information is given.

digest

digest finds the positions where a specified proteolytic enzyme or reagent might cut a peptide sequence.

Here is a sample session with *digest*:

```
% digest
Input sequence: sw:opsd_human
Enzymes and Reagents
         1 : Trypsin
         2 : Lys-C
         3 : Arg-C
         4 : Asp-N
         5 : V8-bicarb
         6 : V8-phosph
         7 : Chymotrypsin
         8 : CNBr
Select number [1]:
Output file [opsd_human.digest]:
```

Mandatory qualifiers:

[-sequencea] (sequence)
: Sequence USA.

-menu (menu)
: Select number.

[-outfile] (report)
: Output report filename.

Advanced qualifiers:

-unfavoured (boolean)
: Trypsin will not normally cut after a K if it is followed by another K or a P. Specifying this shows those cuts as well as the favoured ones.

-aadata (string)
: Molecular weight data for amino acids.

-overlap (boolean)
: Used for partial digestion. Shows all cuts from favoured cut sites plus 1–3, 2–4, 3–5, etc. Overlaps are fragments with exactly one potential cut site within them (e.g., 2–5 is invalid).

-allpartials (boolean)
: Similar to overlap, but fragments containing more than one potential cut site are included.

distmat

distmat calculates the evolutionary distances between every pair of sequences in a multiple alignment. The sequences must be aligned before running this program.

Here is a sample session with *distmat*:

```
% distmat pax.align
Creates a distance matrix from multiple alignments
Multiple substitution correction methods for proteins
```

```
            0 : Uncorrected
            1 : Jukes-Cantor
            2 : Kimura Protein
    Method to use [0]: 2
    Output file [outfile.distmat]:
```

Mandatory qualifiers (bold if not always prompted):

[-msf] (seqset)
: File containing a sequence alignment.

-nucmethod *(menu)*
: Multiple substitution correction methods for nucleotides.

-protmethod *(menu)*
: Multiple substitution correction methods for proteins.

[-outf] (outfile)
: A name for the distance matrix.

Optional qualifiers (bold if not always prompted):

-ambiguous *(boolean)*
: Use the ambiguous codes in the calculation of the Jukes-Cantor method, or if the sequences are proteins.

-gapweight *(float)*
: Weight gaps in the uncorrected (nucleotide) and Jukes-Cantor distance methods.

-position *(integer)*
: Choose base positions to analyse in each codon e.g., 123 (all bases), 12 (the first two bases), 1, 2, or 3 individual bases.

-calculatea *(boolean)*
: Forces the calculation of the a-parameter in the Jin-Nei Gamma distance calculation. Otherwise, the default is 1.0 (see `-parametera` option).

-parametera *(float)*
: User-defined parameter to use in the Jin-Nei Gamma distance calculation. The suggested value (and default) is 1.0.

domainer

domainer reads protein coordinate files and writes domains coordinate files.

Here is a sample session with *domainer*:

```
% domainer
Build domain coordinate files
Name of scop file for input (embl-like format) [Escop.dat]: /data/scop/
Escop.dat
Location of coordinate files for input (embl-like format) [./]: /data/cpdb/
Location of coordinate files for output (embl-like format) [./]:
Extension of coordinate files (embl-like format) [.pxyz]:
Location of coordinate files for output (pdb format) [./]:
Extension of coordinate files (pdb format) [.ent]:
Name of log file for the embl-like format build [domainer.log1]: log.1
Name of log file for the pdb format build [domainer.log2]: log.2
D3SDHA_
D3SDHB_
```

```
D3HBIA_
D3HBIB_
D4SDHA_
D4SDHB_
D4HBIA_
D4HBIB_
D5HBIA_
D5HBIB_
D7HBIA_
D7HBIB_
```

Mandatory qualifiers:

-scop (infile)
    Name of scop classification file (EMBL format input).

[-cpdb] (string)
    Location of protein coordinate files (EMBL format input).

[-cpdbextn] (string)
    Extension of coordinate files (EMBL format).

[-cpdbscop] (string)
    Location of domain coordinate files (EMBL format output)

[-pdbscop] (string)
    Location of domain coordinate files (PDB format output).

[-pdbextn] (string)
    Extension of coordinate files (PDB format).

-cpdberrf (outfile)
    Name of log file for the EMBL format build.

-pdberrf (outfile)
    Name of log file for the PDB format build.

dotmatcher

dotmatcher displays a thresholded dotplot of two sequences.

Here is a sample session with *dotmatcher*:

```
% dotmatcher sw:hba_human sw:hbb_human
```

Mandatory qualifiers (bold if not always prompted):

[-sequencea] (sequence)
    Sequence USA.

[-sequenceb] (sequence)
    Sequence USA.

-data *(boolean)*
    Output the match data to a file instead of plotting it.

-graph *(graph)*
    Graph type.

-xygraph *(xygraph)*
    Graph type.

-outfile *(outfile)*
    Display as data.

Optional qualifiers:

-windowsize (integer)
: Window size over which to test threshold.

-threshold (integer)
: Threshold.

-matrixfile (matrix)
: This is the scoring matrix file used when comparing sequences. By default, the file is *EBLOSUM62* (for proteins) or *EDNAFULL* (for nucleic sequences). These files are in the *data* directory of the EMBOSS installation.

Advanced qualifiers:

-stretch (boolean)
: Display a nonproportional graph.

dotpath

dotpath displays a non-overlapping wordmatch dotplot of two sequences. *dotpath* is very similar to the program *dottup*, which looks for places where words (tuples) of a specified length have an exact match in both sequences and draws a diagonal line over the position of those words.

Here is a sample session with *dotpath*:

```
% dotpath embl:AF129756 embl:AP000504 -word 20
Displays a non-overlapping wordmatch dotplot of two sequences
Graph type [x11]:
```

Mandatory qualifiers (bold if not always prompted):

[-sequencea] (sequence)
: Sequence USA.

[-sequenceb] (sequence)
: Sequence USA.

-wordsize (integer)
: Word size.

-graph *(graph)*
: Graph type.

-outfile *(outfile)*
: Output filename.

Optional qualifiers:

-overlaps (boolean)
: Displays the overlapping matches (in red) and the minimal set of non-overlapping matches.

-[no]boxit (boolean)
: Draw a box around dotplot.

Advanced qualifiers:

-data (boolean)
: Output the match data to a file instead of plotting it.

dottup

dottup looks for places where words (tuples) of a specified length have an exact match in both sequences and draws a diagonal line over the position of these words.

Here is a sample session with *dottup*:

```
% dottup embl:eclac embl:eclaci -wordsize=6 -gtitle="eclaci vs eclac"
```

Mandatory qualifiers (bold if not always prompted):

[-sequencea] (sequence)
Sequence USA.

[-sequenceb] (sequence)
Sequence USA.

-wordsize (integer)
Word size.

-data *(boolean)*
Output the match data to a file instead of plotting it.

-graph *(graph)*
Graph type.

-xygraph *(xygraph)*
Graph type.

-outfile *(outfile)*
Output filename.

Optional qualifiers (bold if not always prompted):

-**[no]boxit** *(boolean)*
Draw a box around dotplot.

Advanced qualifiers:

-stretch (boolean)
Use nonproportional axes.

dreg

dreg searches for matches of a regular expression to a nucleic acid sequence.

Here is a sample session with *dreg*:

```
% dreg
Input sequence: embl:paamir
Output file [paamir.dreg]:
Regular expression pattern: ggtacc
```

Mandatory qualifiers:

[-sequence] (seqall)
Sequence database USA.

[-pattern] (regexp)
Regular expression pattern.

[-outfile] (outfile)
Output filename.

einverted

einverted looks for inverted repeats (stem loops) in a nucleotide sequence.

Here is a sample session with *einverted*:

```
% einverted
Input sequence: embl:hsts1
Output file [hsts1.inv]:
Gap penalty [12]:
Minimum score threshold [50]:
Match score [3]:
Mismatch score [-4]:
```

Mandatory qualifiers:

[-sequence] (sequence)
    Sequence USA.

-gap (integer)
    Gap penalty.

-threshold (integer)
    Minimum score threshold.

-match (integer)
    Match score.

-mismatch (integer)
    Mismatch score.

[-outfile] (outfile)
    Output filename.

embossdata

embossdata is a utility that indicates what directories can hold EMBOSS data files and displays the names of these files.

Display the directories searched for EMBOSS data files:

```
% embossdata
```

Display the names of data file in all of the possible data directories:

```
% embossdata -showall
```

Display the directories which contain a particular EMBOSS data file:

```
% embossdata -filename EPAM60
```

Make a copy of an EMBOSS data file in the current directory:

```
% embossdata -fetch -filename Epepcoil.dat
```

Mandatory qualifiers (bold if not always prompted):

-filename *(string)*
    This specifies the name of the file to fetch into the current directory or search for in all EMBOSS-searchable directories. The name of the file is not altered when it is fetched.

Optional qualifiers (bold if not always prompted):

-showall (boolean)
: Show all potential EMBOSS data files.

-fetch *(boolean)*
: Fetch a data file.

-outf (outfile)
: This specifies the filename that results of a file search are written to. By default, these results are written to the screen (`stdout`).

Advanced qualifiers:

-reject (selection)
: This specifies subdirectories of the EMBOSS data directory to ignore when displaying data directories.

EMBOSS

embossversion

embossversion writes out the current version of the EMBOSS package.

Here is a sample session with *embossversion*:

```
% embossversion
Writes the current EMBOSS version number
1.13.0
```

Optional qualifiers:

-outfile (outfile)
: Output file.

emma

emma calculates the multiple alignment of nucleic acid or protein sequences according to the method of J.D. Thompson, D.C. Higgins, and T.J.Gibson. This is an interface to the ClustalW distribution.

Here is an example session with *emma*:

```
% emma
Input sequence: globins.fasta
Output sequence [hbahum.aln]:
Output file [hbahum.dnd]:
..clustalw17 -infile=5345A -outfile=5345B -align -type=protein ...

 CLUSTAL W (1.74) Multiple Sequence Alignments

Sequence type explicitly set to Protein
Sequence format is Pearson
Sequence 1: hbahum          141 aa
Sequence 2: hbbhum          146 aa
Sequence 3: hbghum          146 aa
Sequence 4: hbhagf          148 aa
Sequence 5: hbrlam          149 aa
Sequence 6: mycrhi          151 aa
Sequence 7: myohum          153 aa
Start of Pairwise alignments
```

```
Aligning...
Sequences (1:2) Aligned. Score:  41
Sequences (1:3) Aligned. Score:  39
Sequences (1:4) Aligned. Score:  21
Sequences (1:5) Aligned. Score:  27
Sequences (1:6) Aligned. Score:  13
Sequences (1:7) Aligned. Score:  26
Sequences (2:3) Aligned. Score:  73
Sequences (2:4) Aligned. Score:  19
Sequences (2:5) Aligned. Score:  19
Sequences (2:6) Aligned. Score:  15
Sequences (2:7) Aligned. Score:  24
Sequences (3:4) Aligned. Score:  21
Sequences (3:5) Aligned. Score:  21
Sequences (3:6) Aligned. Score:  15
Sequences (3:7) Aligned. Score:  23
Sequences (4:5) Aligned. Score:  41
Sequences (4:6) Aligned. Score:  12
Sequences (4:7) Aligned. Score:  16
Sequences (5:6) Aligned. Score:  17
Sequences (5:7) Aligned. Score:  18
Sequences (6:7) Aligned. Score:  11
Guide tree        file created:   [5345C]
Start of Multiple Alignment
There are 6 groups
Aligning...
Group 1: Sequences:   2       Score:883
Group 2: Sequences:   2       Score:2344
Group 3: Sequences:   3       Score:934
Group 4:                      Delayed
Group 5: Sequences:   5       Score:950
Group 6:                      Delayed
Sequence:7      Score:1046
Sequence:6      Score:986
Alignment Score 1746
GCG-Alignment file created      [5345B]
```

Mandatory qualifiers:

[-inseqs] (seqall)
: Sequence database USA.

[-outseq] (seqoutset)
: The sequence alignment output filename.

[-dendoutfile] (outfile)
: The dendogram output filename.

Optional qualifiers (bold if not always prompted):

-onlydend (boolean)
: Produce only a dendrogram file.

-dend *(boolean)*
: Select if you want to perform alignment using an old dendrogram.

-dendfile *(string)*
: Name of the old dendrogram file.

-insist (boolean)
Insist that the sequence type be changed to protein.

-slowfast (menu)
A distance is calculated between every pair of sequences, then these distances are used to construct a dendrogram that guides the final multiple alignment. The scores are calculated from separate pairwise alignments. These can be calculated using 2 methods: dynamic programming (slow but accurate), or by the method of Wilbur and Lipman (extremely fast but approximate). The slow but accurate method is fine for short sequences, but will be extremely slow for many (e.g., greater than100) long (e.g., greater than 1000 residue) sequences.

-pwgapc *(float)*
The penalty for opening a gap in the pairwise alignments.

-pwgapv *(float)*
The penalty for extending a gap by 1 residue in the pairwise alignments.

-pwmatrix *(menu)*
A scoring table that describes the similarity of each amino acid to one another. There are three built-in series of weight matrices offered. Each consists of several matrixes that work differently at different evolutionary distances. For details, read the documentation. Crudely, we store several matrices in memory, spanning the full range of amino acid distance (from almost identical sequences to highly divergent ones). For very similar sequences, it is best to use a strict weight matrix which gives a high score only to identities and the most favoured conservative substitutions. For more divergent sequences, it is appropriate to use "softer" matrixes that give a high score to many other frequent substitutions.

1. BLOSUM (Henikoff). These matrixes appear to be the best available for carrying out data base similarity (homology searches). The matrixes used are: Blosum80, 62, 45 and 30.
2. PAM (Dayhoff). These have been extremely widely used since the late 1970s. We use the PAM 120, 160, 250 and 350 matrixes.
3. GONNET. These matrices were derived using almost the same procedure as the Dayhoff one (above) but are much more up to date and are based on a far larger data set. They appear to be more sensitive than the Dayhoff series. We use the GONNET 40, 80, 120, 160, 250 and 350 matrixes. We also supply an identity matrix which gives a score of 1.0 to two identical amino acids and a score of zero otherwise. This matrix is not very useful.

-pwdnamatrix *(menu)*
A scoring table that describes the scores assigned to matches and mismatches (including IUB ambiguity codes).

-pairwisedata *(string)*
Filename of user pairwise matrix.

-ktup *(integer)*
This is the size of the exact matching fragment. Increase for speed (maximum is 2 for proteins, 4 for DNA); decrease for sensitivity. For longer sequences (e.g., greater than1000 residues), you may need to increase the default.

-gapw *(integer)*
A penalty for each gap in the fast alignments. It has little affect on the speed or sensitivity except in the case of extreme values.

-topdiags (*integer)*
: The number of k-tuple matches on each diagonal (in an imaginary dot matrix plot) is calculated. Only the best ones (those with the most matches) are used in the alignment. Decrease for speed; increase for sensitivity.

-window (*integer)*
: This is the number of diagonals around each of the best diagonals that will be used. Decrease for speed; increase for sensitivity.

-nopercent (*boolean)*
: Fast pairwise alignment: similarity scores: suppresses percentage score.

-matrix (*menu)*
: This gives a menu where you are offered a choice of weight matrices. The default for proteins is the PAM series derived by Gonnet and colleagues. Note that a series is used! The matrix used is dependent upon the similarity of the sequences to be aligned at this alignment step. Different matrixes work differently at each evolutionary distance. There are three built-in series of weight matrixes offered. Each consists of several matrixes that work differently at different evolutionary distances. For details, read the documentation. Crudely, we store several matrices in memory, spanning the full range of amino acid distance (from almost identical sequences to highly divergent ones). For very similar sequences, it is best to use a strict weight matrix which gives a high score only to identities and the most favoured conservative substitutions. For more divergent sequences, it is appropriate to use "softer" matrices that give a high score to many other frequent substitutions.

1. BLOSUM (Henikoff). These matrixes appear to be the best available for carrying out data base similarity (homology searches). The matrixes used are: Blosum 80, 62, 45 and 30.
2. PAM (Dayhoff). These have been widely used since the late 1970s. We use the PAM 120, 160, 250 and 350 matrixes.
3. GONNET. These matrices were derived using almost the same procedure as Dayhoff (above), but are much more up to date and are based on a much larger data set. They appear to be more sensitive than the Dayhoff series. We use the GONNET 40, 80, 120, 160, 250 and 350 matrixes. We also supply an identity matrix which gives a score of 1.0 to two identical amino acids and a score of zero otherwise. This matrix is not very useful. Alternatively, you can read in your own (just one matrix, not a series).

-dnamatrix (*menu)*
: Provides a menu containing a submenu in which a single matrix (not a series) can be selected.

-mamatrix (*string)*
: Filename of multiple user alignment matrix.

-gapc (float)
: Penalty for opening a gap in the alignment. Increasing the gap opening penalty will make gaps less frequent.

-gapv (float)
: Penalty for extending a gap by 1 residue. Increasing the gap extension penalty makes gaps shorter. Terminal gaps are not penalized.

-[no]endgaps (boolean)

"End gap separation" treats end gaps as internal gaps for the purposes of avoiding gaps that are too close (set by "gap separation distance"). If you turn this off, end gaps will be ignored. This is useful when you want to align fragments where the end gaps are not biologically meaningful.

-gapdist (integer)

"Gap separation distance" tries to decrease the chances of gaps being too close. Gaps that are less than this distance apart are penalized more than other gaps. This does not prevent close gaps; it only makes them less frequent, resulting in alignments that have a blocklike appearance.

-norgap *(boolean)*

"Residue specific penalties" are amino acid-specific gap penalties that reduce or increase the gap opening penalties at each position in the alignment or sequence. As an example, positions that are rich in glycine are more likely to have an adjacent gap than positions that are rich in valine.

-hgapres *(string)*

A set of the residues considered hydrophilic. It is used when introducing Hydrophilic gap penalties.

-nohgap *(boolean)*

"Hydrophilic gap penalties" are used to increase the chances of a gap within a run (5 or more residues) of hydrophilic amino acids; these are likely to be loop or random coil regions where gaps are more common. The residues that are considered hydrophilic are set by `-hgapres`.

-maxdiv (integer)

This switch delays the alignment of the most distantly related sequences until after the most closely related sequences are aligned. The setting shows the percent identity level required to delay the addition of a sequence.

Advanced qualifiers:

-prot (boolean)

Do not change this value.

---

emowse

emowse will search a protein database for matches with the mass spectrometry data.

Here is a sample session with *emowse*:

```
% emowse
Protein identification by mass spectrometry
Input sequence(s): sw:*
Input file: test.mowse
Whole sequence molwt [0]:
Output file [100k_rat.emowse]:
```

Mandatory qualifiers:

[-sequences] (seqall)

Sequence database USA.

[-infile] (infile)

Name of molecular weight data file.

-weight (integer)
   Whole sequence molecular weight.

-outfile (outfile)
   Output filename.

Advanced qualifiers:

-enzyme (menu)
   Enzyme or reagent.

-aadata (string)
   Molecular weight data for amino acids.

-pcrange (integer)
   Allowed whole sequence weight variability.

-frequencies (string)
   Frequencies file.

-tolerance (float)
   Float value

-partials (float)
   Partials factor.

entret

entret reads a sequence from a database or a file and writes the complete sequence entry to a text file.

Here is a sample session with *entret*:

```
% entret embl:hsfau
Reads and writes (returns) flat file entries
Output file [hsfau.entret]:
```

Mandatory qualifiers:

[-sequence] (seqall)
   Sequence database USA.

[-outfile] (outfile)
   Output filename.

Advanced qualifiers:

-firstonly (boolean)
   Read one sequence and stop.

eprimer3

eprimer3 is an interface to the *primer3* program from the Whitehead Institute. The Whitehead program must be set up and on the path in order for *eprimer3* to find and run it. *eprimer3* picks primers for PCR reactions, considering as criteria:

- Oligonucleotide melting temperature, size, GC content, and primer-dimer possibilities.
- PCR product size.
- Positional constraints within the source sequence.
- Other miscellaneous constraints.

Here is a sample session with *eprimer3*:

```
% eprimer3 em:hsfau1 hsfau.eprimer3 -explain
```

Mandatory qualifiers:

[-sequence] (seqall)
: The sequence from which to choose primers. The sequence must be presented 5' to 3'.

[-outfile] (outfile)
: Output filename.

Optional qualifiers (bold if not always prompted):

-task (menu)
: Tell *eprimer3* what task to perform. Legal values are:

0. Pick PCR primer.
1. Pick PCR primers and hybridization probe.
2. Pick forward primer only.
3. Pick reverse primer only.
4. Pick hybridization probe only.

The tasks should be self explanatory. Briefly, an internal oligo is intended to be used as a hybridization probe (hyb probe) to detect the PCR product after amplification.

-numreturn (integer)
: The maximum number of primer pairs to return. Primer pairs returned are sorted by their quality, in other words, by the value of the objective function (where a lower number indicates a better primer pair). Note that setting this parameter to a large value will increase running time.

-includedregion (range)
: A subregion of the given sequence from which to pick primers. For example, the first dozen or so bases of a sequence are often vector and should be excluded from consideration. The value for this parameter has the form `start,end` where `start` is the index of the first base to consider and `end` is the last in the primer-picking region.

-target (range)
: If one or more Targets are specified, a legal primer pair must flank at least one of them. A Target might be a simple sequence repeat site (for example, a CA repeat) or a single-base-pair polymorphism. The value should be a space-separated list of `start,end` pairs where `start` is the index of the first base of a Target, and `end` is the last. E.g., `50,51` requires primers to surround the 2 bases at positions 50 and 51.

-excludedregion (range)
: Primer oligos may not overlap any region specified in this tag. The associated value must be a space-separated list of `start,end` pairs where `start` is the index of the first base of the excluded region, and `end` is the last. This tag is useful for tasks such as excluding regions of low sequence quality or excluding regions containing repetitive elements such as ALUs or LINEs. E.g., `401,407 68,70` forbids selection of primers in the 7 bases starting at 401 and the 3 bases at 68.

-forwardinput (string)
: The sequence of a forward primer to check, then use to design reverse primers and optional internal oligos. Must be a substring of SEQUENCE.

-reverseinput (string)
: The sequence of a reverse primer to check and use to design forward primers and optional internal oligos. Must be a substring of the reverse strand of SEQUENCE.

-gcclamp *(integer)*
: Require the specified number of consecutive Gs and Cs at the 3' end of both the forward and reverse primer. (This parameter has no effect on the internal oligo if one is requested.)

-osize *(integer)*
: Optimum length (in bases) of a primer oligo. EPrimer3 attempts to pick primers close to this length.

-minsize *(integer)*
: Minimum acceptable length of a primer. Must be greater than 0 and less than or equal to MAX-SIZE.

-maxsize *(integer)*
: Maximum acceptable length (in bases) of a primer. Currently, this parameter cannot be larger than 35. This limit is governed by the maximum oligo size for which EPrimer3's melting-temperature is valid.

-otm *(float)*
: Optimum melting temperature (Celsius) for a primer oligo. EPrimer3 will try to pick primers with melting temperatures close to this temperature. The oligo melting temperature formula in EPrimer3 is given in Rychlik, Spencer and Rhoads, *Nucleic Acids Research*, Volume 18:12, pages 6409-6412 and Breslauer, Frank, Bloeker and Marky, *Proceedings of the National Academy of Sciences USA*, vol 83, pp 3746-3750. Please refer to the former paper for background discussion.

-mintm *(float)*
: Minimum acceptable melting temperature (Celsius) for a primer oligo.

-maxtm *(float)*
: Maximum acceptable melting temperature (Celsius) for a primer oligo.

-maxdifftm *(float)*
: Maximum acceptable (unsigned) difference between the melting temperatures of the forward and reverse primers.

-ogcpercent *(float)*
: Primer optimum GC percent.

-mingc *(float)*
: Minimum allowable percentage of Gs and Cs in any primer.

-maxgc *(float)*
: Maximum allowable percentage of Gs and Cs in any primer generated by Primer.

-saltconc *(float)*
: The millimolar concentration of salt (usually KCl) in the PCR. EPrimer3 uses this argument to calculate oligo melting temperatures.

-dnaconc *(float)*
: The nanomolar concentration of annealing oligos in the PCR. EPrimer3 uses this argument to calculate oligo melting temperatures. The default (50nM) works well with the standard protocol used at the Whitehead/MIT Center for Genome

Research—0.5 microliters of 20 micromolar concentration for each primer oligo in a 20 microliter reaction with 10 nanograms template, 0.025 units/microliter Taq polymerase in 0.1 mM each dNTP, 1.5mM $MgCl_2$, 50mM KCl, 10mM Tris-HCl (pH 9.3) using 35 cycles with an annealing temperature of 56 degrees Celsius. This parameter corresponds to "c" in Rychlik, Spencer and Rhoads' equation (ii) (*Nucleic Acids Research*, vol 18, num 12) where a suitable value (for a lower initial concentration of template) is "empirically determined". The value of this parameter is less than the actual concentration of oligos in the reaction because it is the concentration of annealing oligos, which in turn depends on the amount of template (including PCR product) in a given cycle. This concentration increases a great deal during a PCR; fortunately PCR seems quite robust for a variety of oligo melting temperatures.

-maxpolyx (*integer*)
: The maximum allowable length of a mononucleotide repeat in a primer, for example `AAAAAA`.

-productosize (*integer*)
: The optimum size for the PCR product. `0` indicates that there is no optimum product size.

-productsizerange (*range*)
: The associated values specify the lengths of the product that the user wants the primers to create, and is a space separated list of elements of the form *x-y* where this pair is a legal range of lengths for the product. For example, if you want PCR products to be between 100 to 150 bases (inclusive), set this parameter to `100-150`. If you desire PCR products in either the range from 100 to 150 bases or in the range from 200 to 250 bases, set this parameter to `100-150 200-250`. EPrimer3 favors ranges to the left side of the parameter string. EPrimer3 will return legal primers pairs in the first range regardless the value of the objective function for these pairs. EPrimer3 returns primers in a subsequent range only if there is an insufficient number of primers in the first range.

-productotm (*float*)
: The optimum melting temperature for the PCR product. `0` indicates that there is no optimum temperature.

-productmintm (*float*)
: The minimum allowed melting temperature of the amplicon. Please see the documentation on the maximum melting temperature of the product for details.

-productmaxtm (*float*)
: The maximum allowed melting temperature of the amplicon. Product Tm is calculated using the formula from Bolton and McCarthy, *Proceedings of the National Academy of Sciences* 84:1390 (1962) as presented in Sambrook, Fritsch and Maniatis, *Molecular Cloning*, p 11.46 (1989, CSHL Press). T_m = 81.5 + 16. 6(log_{10}([Na^+])) + .41*(%GC) - 600/length where [Na^+} is the molar sodium concentration, (%GC) is the percent of Gs and Cs in the sequence, and length is the length of the sequence. A similar formula is used by the prime primer selection program in GCG (*http://www.gcg.com*), which instead uses 675.0/length in the last term (after F. Baldino, Jr, M.-F. Chesselet, and M.E. Lewis, *Methods in Enzymology* 168:766 (1989) eqn (1) on page 766 without the mismatch and formamide terms). The formulas here and in Baldino et al. assume Na^+ rather than K^+. According to J.G. Wetmur, Critical Reviews in Biochemistry and and Molecular Biology 26:227 (1991) 50 mM K^+ should be equivalent in these

formulae to .2 M Na+. EPrimer3 uses the same salt concentration value for calculating both the primer melting temperature and the oligo melting temperature. If you plan to later use the PCR product for hybridization, this behavior will not give you the Tm under hybridization conditions.

-oligoexcludedregion *(range)*
: Middle oligos may not overlap any region specified by this tag. The associated value must be a space-separated list of `start,end` pairs, where `start` is the index of the first base of an excluded region, and *end* is the last. Often one would make Target regions excluded regions for internal oligos.

-oligoinput *(string)*
: The sequence of an internal oligo to check, then use to design forward and reverse primers. Must be a substring of SEQUENCE.

-oligoosize *(integer)*
: Optimum length (in bases) of an internal oligo. EPrimer3 will attempt to pick primers close to this length.

-oligominsize *(integer)*
: Minimum acceptable length of an internal oligo. Must be greater than `0` and less than or equal to `INTERNAL-OLIGO-MAX-SIZE`.

-oligomaxsize *(integer)*
: Maximum acceptable length (in bases) of an internal oligo. Currently, this parameter cannot be larger than `35`. This limit is governed by maximum oligo size for which EPrimer3's melting temperature is valid.

-oligootm *(float)*
: Optimum melting temperature (Celsius) for an internal oligo. EPrimer3 tries to pick oligos with melting temperatures close to this temperature. The oligo melting temperature formula in EPrimer3 is that given in Rychlik, Spencer and Rhoads, *Nucleic Acids Research*, vol 18, num 12, pp 6409-6412 and Breslauer, Frank, Bloeker and Marky, *Procedures of the National Academy of Sciences USA*, vol 83, pp 3746-3750. Please refer to the former paper for background discussion.

-oligomintm *(float)*
: Minimum acceptable melting temperature (Celsius) for an internal oligo.

-oligomaxtm *(float)*
: Maximum acceptable melting temperature (Celsius) for an internal oligo.

-oligoogcpercent *(float)*
: Internal oligo optimum GC percent.

-oligomingc *(float)*
: Minimum allowable percentage of Gs and Cs in an internal oligo.

-oligomaxgc *(float)*
: Maximum allowable percentage of Gs and Cs in any internal oligo generated by Primer.

-oligosaltconc *(float)*
: The millimolar concentration of salt (usually KCl) in the hybridization. EPrimer3 uses this argument to calculate internal oligo melting temperatures.

-oligodnaconc *(float)*
: The nanomolar concentration of annealing internal oligo in the hybridization.

-oligoselfany (*float)*

The maximum allowable local alignment score when testing an internal oligo for (local) self-complementarity. Local self-complementarity is taken to predict the tendency of oligos to anneal to themselves. The scoring system gives 1.00 for complementary bases, -0.25 for a match of any base (or N) with an N, -1.00 for a mismatch, and -2.00 for a gap. Only single-base-pair gaps are allowed. For example, the alignment:

```
5' ATCGNA 3'
   || | |
3' TA-CGT 5'
```

is allowed (and yields a score of 1.75), but the alignment:

```
5' ATCCGNA 3'
   ||  | |
3' TA--CGT 5'
```

is not considered. Scores are non-negative, and a score of 0.00 indicates that there is no reasonable local alignment between two oligos.

-oligoselfend (*float)*

The maximum allowable 3'-anchored global alignment score when testing a single oligo for self-complementarity. The scoring system is as for the Maximum Complementarity argument. In the examples above the scores are 7.00 and 6.00 respectively. Scores are non-negative, and a score of 0.00 indicates that there is no reasonable 3'-anchored global alignment between two oligos. In order to estimate 3'-anchored global alignments for candidate oligos, Primer assumes that the sequence from which to choose oligos is presented 5' to 3'. `INTERNAL-OLIGO-SELF-END` is meaningless when applied to internal oligos used for hybridization-based detection, since primer-dimer will not occur. We recommend that `INTERNAL-OLIGO-SELF-END` be set at least as high as `INTERNAL-OLIGO-SELF-ANY`.

-oligomaxpolyx (*integer)*

The maximum allowable length of an internal oligo mononucleotide repeat. For example, `AAAAAA`.

Advanced qualifiers:

-explainflag (boolean)

If this flag is nonzero, produce `LEFT-EXPLAIN`, `RIGHT-EXPLAIN`, and `INTERNAL-OLIGO-EXPLAIN` output tags, which provide information on the number of oligos and primer pairs examined by EPrimer3. These tags also provide statistics on the number discarded and the reasons for these discards.

-fileflag (boolean)

If the associated value is nonzero, EPrimer3 creates two output files for each input SEQUENCE. File *sequence-id.for* lists all acceptable forward primers for *sequence-id*, and *sequence-id.rev* lists all acceptable reverse primers for *sequence-id*, where *sequence-id* is the value of the `SEQUENCE-ID` tag (which must be supplied). In addition, if the input tag TASK is 1 or 4, EPrimer3 produces a file *sequence-id.int*, which lists all acceptable internal oligos.

-firstbaseindex (integer)

This parameter is the index of the first base in the input sequence. For input and output using 1-based indexing (the method GenBank uses), set this parameter to 1. For input and output using 0-based indexing, set this parameter to `0`. (This parameter also affects the indexes in the contents of the files produced when the primer file flag is set.)

EMBOSS

-pickanyway (boolean)

If true, pick a primer pair—even if LEFT-INPUT, RIGHT-INPUT, or INTERNAL-OLIGO-INPUT violate specific constraints.

-mispriminglibrary (infile)

The name of a file containing a nucleotide sequence library of sequences to avoid amplifying. For example, repetitive sequences or the sequences of genes in a gene family that should not be amplified are often put into this file. The file must be in (a slightly restricted) FASTA format (W. B. Pearson and D.J. Lipman, *PNAS* 85:8 pp 2444-2448 [1988]); we briefly discuss the organization of this file below. If this parameter is specified, EPrimer3 locally aligns each candidate primer against each library sequence and rejects those primers for which the local alignment score times a specified weight (see below) exceeds MAX-MISPRIMING. (The maximum value of the weight is arbitrarily set to 100.0.) Each sequence entry in the FASTA format file must begin with an id line that starts with ">". The contents of the id line is slightly restricted in that EPrimer3 parses everything after any optional asterisk ("*") as a floating point number to use as the weight mentioned previously. If the id line contains no asterisk, the weight defaults to 1.0. The alignment scoring system used is the same as for calculating complementarity among oligos (e.g., SELF-ANY). The remainder of an entry contains the sequence as lines following the id line up until a line starting with ">" or the end of the file. Whitespace and newlines are ignored. Characters "A", "T", "G", "C'", "a", "t", "g", "c" are retained. Any other character is converted to "N" (meaning that any IUB/IUPAC codes for ambiguous bases are converted to "N"). There are no restrictions on line length. An empty value for this parameter indicates that no repeat library should be used.

-maxmispriming (float)

The maximum allowed weighted similarity with any sequence in MISPRIMING-LIBRARY.

-pairmaxmispriming (float)

The maximum allowed sum of weighted similarities of a primer pair (one similarity for each primer) with any single sequence in MISPRIMING-LIBRARY.

-numnsaccepted (integer)

Maximum number of unknown bases N allowable in any primer.

-selfany (float)

The maximum allowable local alignment score when testing a single primer for (local) self-complementarity, and the maximumallowable local alignment score when testing for complementarity between forward and reverse primers. Local self-complementarity is taken to predict the tendency of primers to anneal to each other without necessarily causing self-priming in the PCR. The scoring system gives 1.00 for complementary bases, -0.25 for a match of any base (or N) with an N, -1.00 for a mismatch, and -2.00 for a gap. Only single-base-pair gaps are allowed. For example, the alignment:

```
5' ATCGNA 3'
...|| | |
3' TA-CGT 5'
```

is allowed (and yields a score of 1.75), but the alignment:

```
5' ATCCGNA 3'
...||  | |
3' TA--CGT 5'
```

is not considered. Scores are non-negative, and a score of 0.00 indicates that there is no reasonable local alignment between two oligos.

-selfend (float)

The maximum allowable 3'-anchored global alignment score when testing a single primer for self-complementarity, and the maximum allowable 3'-anchored global alignment score when testing for complementarity between forward and reverse primers. The 3'-anchored global alignment score is taken to predict the likelihood of PCR-priming primer-dimers, for example:

```
5' ATGCCCTAGCTTCCGGATG 3'
.............||| |||||..........
3'            AAGTCCTACATTTAGCCTAGT 5'
```

or:

```
5' AGGCTATGGGCCTCGCGA 3'
...............||||||............
3'               AGCGCTCCGGGTATCGGA 5'
```

The scoring system is as for the Maximum Complementarity argument. In the previous examples, the scores are 7.00 and 6.00, respectively. Scores are non-negative, and a score of 0.00 indicates that there is no reasonable 3'-anchored global alignment between two oligos. In order to estimate 3'-anchored global alignments for candidate primers and primer pairs, Primer assumes that the sequence from which to choose primers is presented 5' to 3'. It is nonsensical to provide a larger value for this parameter than that given for the Maximum (local) Complementarity parameter because the score of a local alignment will always be at least as great as the score of a global alignment.

-maxendstability (float)

The maximum stability for the five 3' bases of a forward or reverse primer. Bigger numbers mean more stable 3' ends. The value is the maximum delta G for duplex disruption for the five 3' bases as calculated using the nearest neighbor parameters published in Breslauer, Frank, Bloeker and Marky, *Proceedings of the National Academy of Sciences,* vol 83, pp 3746-3750. EPrimer3 uses a completely permissive default value for backward compatibility (which we may change in the next release). Rychlik recommends a maximum value of 9 (Wojciech Rychlik, "Selection of Primers for Polymerase Chain Reaction" in BA White, Ed., *Methods in Molecular Biology, Vol. 15: PCR Protocols: Current Methods and Applications*, 1993, pp 31-40, Humana Press, Totowa NJ).

-oligomishyblibrary (infile)

Similar to MISPRIMING-LIBRARY, except that the event we seek to avoid is hybridization of the internal oligo to sequences in this library rather than priming from them. The file must be in (a slightly restricted) FASTA format (W. B. Pearson and D.J. Lipman, *Proceedings of the National Academy fo Sciences* 85:8 pp 2444-2448 [1988]); we briefly discuss the organization of this file below. If this parameter is specified, EPrimer3 locally aligns each candidate oligo against each library sequence and rejects those primers for which the local alignment score times a specified weight exceeds `INTERNAL-OLIGO-MAX-MISHYB`. (The maximum value of the weight is arbitrarily set to 12.0.) Each sequence entry in the FASTA-format file must begin with an id line that starts with ">". The contents of the id line is slightly restricted in that EPrimer3 parses everything after any optional asterisk ("*") as a floating point number to use as the weight mentioned above. If the id line contains no asterisk, the weight defaults to 1.0. The alignment scoring system used is the same as for calculating complementarity among oligos (e.g.,

EMBOSS

SELF-ANY). The remainder of an entry contains the sequence as lines following the id line up until a line starting with ">" or the end of the file. Whitespace and newlines are ignored. Characters "A", "T", "G", "C'", "a", "t", "g", and "c" are retained and any other character is converted to "N" (meaning that any IUB / IUPAC codes for ambiguous bases are converted to "N"). There are no restrictions on line length. An empty value for this parameter indicates that no library should be used.

-oligomaxmishyb (float)
: Similar to *MAX-MISPRIMING*, except this parameter applies to the similarity of candidate internal oligos to the library specified in INTERNAL-OLIGO-MISHYB-LIBRARY.

equicktandem

equicktandem scans a sequence for potential tandem repeats up to a specified size. The results can be used to run *etandem* on the candidate repeat lengths to identify genuine tandem repeats.

Here is a sample session with *equicktandem*. The input sequence is the human herpesvirus tandem repeat:

```
% equicktandem
Input sequence: embl:hhtetra
Output file [hhtetra.qtan]:
Maximum repeat size [600]:
Threshold score [20]:
```

Mandatory qualifiers:

[-sequence] (sequence)
: Sequence USA.

-maxrepeat (integer)
: Maximum repeat size.

-threshold (integer)
: Threshold score.

[-outfile] (report)
: Output report filename.

Advanced qualifiers:

-origfile (outfile)
: Output filename.

est2genome

est2genome will align a set of spliced nucleotide sequences (ESTs, cDNAs, or mRNAs) to an unspliced genomic DNA sequence and insert introns of arbitrary length when needed.

Here is a sample session with *est2genome*:

```
% est2genome
Align EST and genomic DNA sequences
EST sequence(s): embl:hs989235
Genomic sequence: embl:hsnfg9
Output file [hs989235.est2genome]:
```

Mandatory qualifiers:

[-est] (seqall)
EST sequence(s).

[-genome] (sequence)
Genomic sequence.

[-outfile] (outfile)
Output filename.

Optional qualifiers:

-match (integer)
Score for matching two bases.

-mismatch (integer)
Cost for mismatching two bases.

-gappenalty (integer)
Cost for deleting a single base in either sequence, excluding introns.

-intronpenalty (integer)
Cost for an intron, independent of length.

-splicepenalty (integer)
Cost for an intron, independent of length and starting/ending on donor-acceptor sites.

-minscore (integer)
Exclude alignments with scores below this threshold score.

Advanced qualifiers:

-reverse (boolean)
Reverse the orientation of the EST sequence.

-[no]splice (boolean)
Use donor and acceptor splice sites. If you want to ignore donor-acceptor sites, set this to false.

-mode (string)
This determines the comparison mode. The default value is `both`. In this case, both strands of the EST are compared assuming a forward gene direction (ie GT/AG splice sites), and the best comparison redone assuming a reversed (CT/AC) gene splicing direction. The other allowed modes are `forward` (when just the forward strand is searched), and `reverse` (when the reverse strand is searched).

-[no]best (boolean)
You can print out all comparisons (not just the best one) by setting this to false.

-space (float)
For linear-space recursion. If product of sequence lengths divided by 4 exceeds this value, a divide-and-conquer strategy is used to control the memory requirements. Very long sequences can be aligned in this manner. If you have a machine with plenty of memory, you may raise this parameter (but do not exceed the machine's physical RAM).

-shuffle (integer)
Shuffle.

-seed (integer)
Random number seed.

-align (boolean)
: Show the alignment. The alignment includes the first and last 5 bases of each intron, together with the intron width. The direction of splicing is indicated by angle brackets (forward or reverse) or ???? (unknown).

-width (integer)
: Alignment width.

etandem

etandem looks for tandem repeats in a sequence.

Here is a sample session with *etandem*. The input sequence is the human herpes virus tandem repeat:

```
% etandem
Input sequence: embl:hhtetra
Output file [hhtetra.tan]:
Minimum repeat size [10]: 6
Maximum repeat size [6]:
```

Mandatory qualifiers:

[-sequence] (sequence)
: Sequence USA.

-minrepeat (integer)
: Minimum repeat size.

-maxrepeat (integer)
: Maximum repeat size.

[-outfile] (report)
: Output report filename.

Advanced qualifiers:

-threshold (integer)
: Threshold score.

-mismatch (boolean)
: Allow N as a mismatch.

-uniform (boolean)
: Allow uniform consensus.

-origfile (outfile)
: Output filename.

extractfeat

extractfeat is a simple utility for extracting parts of a sequence that have been annotated as a specific type of feature. These subsequences are written to the output sequence file.

Here is a sample session with *extractfeat* to write out the exons of a sequence:

```
% extractfeat embl:hsfau1 -type exon stdout
```

To write out the exons with 10 extra bases at the start and end so that you can inspect the splice sites:

```
% extractfeat embl:hsfau1 -type exon -before 10 -after 10 stdout
```

To write out the 10 bases around the start of all "exon" features in the EMBL database:

```
% extractfeat embl:\*  -type exon -before 5 -after -5 stdout
```

To write out the 7 residues around all phosphorylated residues in SWISS-PROT:

```
% extractfeat sw:\*  -type mod_res -value phosphorylation -before 3 -after -
4 stdout
```

Mandatory qualifiers:

[-sequence] (seqall)
: Sequence database USA.

[-outseq] (seqout)
: Output sequence USA.

Optional qualifiers:

-before (integer)
: If this value is greater than `0`, that number of bases or residues before the feature are included in the extracted sequence. This allows you to see the context of the feature. If this value is negative, the start of the extracted sequence will be this number of bases/residues before the end of the feature. For example, a value of `10` will start the extraction 10 bases/residues before the start of the sequence, and a value of `-10` will start the extraction 10 bases or residues before the end of the feature. The output sequence will be padded with "N" or "X" characters if the sequence starts after the required start of the extraction.

-after (integer)
: If this value is greater than `0`, that number of bases or residues after the feature are included in the extracted sequence. This allows you to see the context of the feature. If this value is negative, the end of the extracted sequence will be this number of bases/residues after the start of the feature. For example, a value of `10` will end the extraction 10 bases/residues after the end of the sequence, and a value of `-10` will end the extraction 10 bases or residues after the start of the feature. The output sequence will be padded with "N" or "X" characters if the sequence ends before the required end of the extraction.

-source (string)
: By default, any feature source in the feature table is shown. You can set this to match any feature source you want to show. The source name is usually either the name of the program that detected the feature, or the feature table (e.g., EMBL) that the feature came from. The source may be wildcarded by using *. If you want to show more than one source, separate their names with the character |, e.g., `gene* | embl`.

-type (string)
: By default, every feature in the feature table is extracted. You can set this to be any feature type you want to extract. See Chapter 2 for a list of the EMBL feature types, and Chapter 3 for a list of the SWISS-PROT feature types. The type may be wildcarded by using *. If you want to extract more than one type, separate their names with the | character. For example:

```
*UTR | intron
```

-sense (integer)
: By default, any feature type in the feature table is extracted. You can set this to match any feature sense you want. `0` matches any sense, `1` matches forward sense, and `-1` matches reverse sense.

-minscore (float)
: If this is greater than or equal to the maximum score, any score is permitted.

-maxscore (float)
: If this is less than or equal to the maximum score, any score is permitted.

-tag (string)
: Tags are the types of extra values that a feature may have. For example, in the EMBL feature table, a CDS type of feature may have the tags `/codon`, `/codon_start`, `/db_xref`, `/EC_number`, `/evidence`, `/exception`, `/function`, `/gene`, `/label`, `/map`, `/note`, `/number`, `/partial`, `/product`, `/protein_id`, `/pseudo`, `/standard_name`, `/translation`, `/transl_except`, `/transl_table`, or `/usedin`. Some of these tags also have values (e.g., `/gene` can have the value of the gene name). By default, any feature tag in the feature table is extracted. You can set this to match any feature tag you want to show. The tag may be wildcarded by using *. If you want to extract more than one tag, separate their names with the | character. For example:

```
gene | label
```

-value (string)
: Tag values are the values associated with a feature tag. Tags are the types of extra values that a feature may have. For example, in the EMBL feature table, a CDS type of feature may have the tags `/codon`, `/codon_start`, `/db_xref`, `/EC_number`, `/evidence`, `/exception`, `/function`, `/gene`, `/label`, `/map`, `/note`, `/number`, `/partial`, `/product`, `/protein_id`, `/pseudo`, `/standard_name`, `/translation`, `/transl_except`, `/transl_table`, or `/usedin`. Some of these tags also have values (e.g., `/gene` can have the value of the gene name). By default, any feature tag in the feature table is extracted. You can set this to match any feature tag value you want to show. The tag may be wildcarded by using *. If you want to extract more than one tag, separate their names with the | character. For example:

```
pax* | 10
```

-join (boolean)
: Some features, such as coding sequence (CDS) and mRNA, are composed of introns concatenated together. There may be other forms of joined sequence, depending on the feature table. If this option is set `TRUE`, any group of these features will be output as a single sequence. If the `before` and `after` qualifiers have been set, only the sequences before the first feature and after the last feature are added.

extractseq

extractseq allows you to specify one or more regions of a sequence to extract subsequences from to build up a contiguous resulting sequence.

Extract the region from position 10 to 20:

```
% extractseq main.seq result.seq -regions '10-20'
```

Extract the regions 10 to 20, 30 to 45, 533 to 537:

```
% extractseq main.seq result2.seq -regions '10-20, 30-45, 533-537'
```

Mandatory qualifiers:

[-sequence] (sequence)
    Sequence USA.

-regions (range)
    Regions to extract. A set of regions is specified by a set of pairs of positions. The positions are integers. They are separated by any non-digit, non-alpha character. Examples of region specifications are: `24-45, 56-78, 1:45, 67=99, 765..888`, or `1,5,8,10,23,45,57,99`.

[-outseq] (seqoutall)
    Output sequence(s) USA.

Optional qualifiers:

-separate (boolean)
    If this is set true, each specified region is written out as a separate sequence. The name of the sequence is created from the name of the original sequence, with underscore characters between the start and end positions of the range. For example, XYZ region 2 to 34 is written as: `XYZ_2_34`.

findkm

findkm takes a file of enzymatic data and plots Michaelis-Menten and Hanes-Woolf plots of the data. From these it calculates the Michaelis-Menten constant (Km) and the maximum velocity (Vmax) of the reaction.

Here is a sample session with *findkm*:

```
%findkm
Calculates Km and Vmax for an enzyme reaction.
Enter name of file containing data: enztest.dat
Display on device [xwin]:
```

Mandatory qualifiers:

[-infile] (infile)
    Enter name of file containing data.

[-outfile] (outfile)
    Output filename.

-graphlb (xygraph)
    Graph type.

Advanced qualifiers:

-[no]plot (boolean)
    S/V versus S.

freak

freak takes one or more sequences as input and a set of bases or residues to search for. It then calculates the frequency of these bases/residues in a window as it moves along the sequence. The frequency is output to a data file or (optionally) plotted.

Here is a sample session with *freak*:

```
% freak embl:hsfau
Residue/base frequency table or plot
Residue letters [gc]:
Output file [hsfau.freak]:
```

Mandatory qualifiers (bold if not always prompted):

[-seqall] (seqall)
Sequence database USA.

-letters (string)
Residue letters.

-graph *(xygraph)*
Graph type.

-outfile *(outfile)*
Output filename.

Optional qualifiers:

-step (integer)
Stepping value.

-window (integer)
Averaging window.

Advanced qualifiers:

-plot (boolean)
Produce graphic.

funky

funky reads clean coordinate files and writes file of protein-heterogen contact data.

Here is a sample session with *funky*:

```
% funky
```

Mandatory qualifiers:

[-prot] (string)
Location of protein coordinate files for input (EMBL-like format).

[-protextn] (string)
Extension of protein coordinate files (EMBL-like format).

[-dom] (string)
Location of domain coordinate files for input (EMBL-like format).

[-domextn] (string)
Extension of domain coordinate files (EMBL-like format).

-dic (infile)
Heterogen groups dictionary name.

-scop (infile)
Name of scop file for input (EMBL-like format).

-vdwf (infile)
Name of data file with van der Waals radii.

-thresh (float)
Threshold contact distance.

Advanced qualifiers:

-outf (outfile)
: Name of output file.

-logf (outfile)
: Name of log file.

---

fuzznuc

fuzznuc uses PROSITE style patterns to search nucleotide sequences.

Here is a sample session with *fuzznuc*:

```
% fuzznuc
Input sequence: embl:hhtetra
Search pattern: AAGCTT
Number of mismatches [0]:
Output file [hhtetra.fuzznuc]:
```

EMBOSS

Mandatory qualifiers:

[-sequence] (seqall)
: Sequence database USA.

-pattern (string)
: The standard IUPAC one-letter codes for the amino acids are used. The letter `x` is used for a position where any amino acid is accepted. Ambiguities are indicated by listing the acceptable amino acids for a given position between square brackets. For example, `[ALT]` stands for Ala, Leu, or Thr. Ambiguities are also indicated by listing the amino acids that are not accepted at a given position in curly brackets. For example, `{AM}` stands for any amino acid except Ala and Met. Each element in a pattern is separated from its neighbor by a dash. (Optional in *fuzznuc*.)

Repetition of an element of the pattern is indicated by following that element with a numerical value or a numerical range between parenthesis. For example, `x(3)` corresponds to `x-x-x`, while `x(2,4)` corresponds to `x-x`, `x-x-x`, or `x-x-x-x`. When a pattern is restricted to either the N- or C-terminal of a sequence, that pattern either starts with a "<" symbol or ends with a ">" symbol, respectively. A period ends the pattern. (Optional in *fuzznuc*.) For example, `[DE](2)HS{P}X(2)PX(2,4)C8`.

-mismatch (integer)
: Number of mismatches.

[-outfile] (report)
: Output report filename.

Advanced qualifiers:

-complement (boolean)
: Search complementary strand.

fuzzpro

fuzzpro uses PROSITE style patterns to search protein sequences.

Here is a sample session with *fuzzpro*:

```
% fuzzpro
Input sequence: sw:*
Search pattern: [FY]-[LIV]-G-[DE]-E-A-Q-x-[RKQ](2)-G
Number of mismatches [0]:
Output file [5h1d_fugru.fuzzpro]:
```

Mandatory qualifiers:

[-sequence] (seqall)
: Sequence database USA.

-pattern (string)
: The standard IUPAC one-letter codes for the amino acids are used. The letter `x` is used for a position where any amino acid is accepted. Ambiguities are indicated by listing the acceptable amino acids for a given position between square brackets. For example, `[ALT]` stands for Ala, Leu, or Thr. Ambiguities are also indicated by listing the amino acids that are not accepted at a given position in curly brackets. For example, `{AM}` stands for any amino acid except Ala and Met. Each element in a pattern is separated from its neighbor by a dash. (Optional in *fuzzpro*.)

  Repetition of an element of the pattern is indicated by following that element with a numerical value or a numerical range between parenthesis. For example, `x(3)` corresponds to `x-x-x`, while `x(2,4)` corresponds to `x-x`, `x-x-x`, or `x-x-x-x`. When a pattern is restricted to either the N- or C-terminal of a sequence, that pattern either starts with a "<" symbol or ends with a ">" symbol, respectively. A period ends the pattern. (Optional in *fuzzpro*.) For example, `[DE](2)HS{P}X(2)PX(2,4)C8`.

-mismatch (integer)
: Number of mismatches.

[-outfile] (report)
: Output report filename.

fuzztran

fuzztran uses PROSITE style protein patterns to search nucleic acid sequences translated in the specified frame(s).

Here is a sample session with *fuzztran*, using all options:

```
% fuzztran -opt
Protein pattern search after translation
Input sequence(s): embl:rnops
Translation frames
          1 : 1
          2 : 2
          3 : 3
          F : Forward three frames
         -1 : -1
         -2 : -2
         -3 : -3
```

```
          R : Reverse three frames
          6 : All six frames
Frame(s) to translate [1]: f
Genetic codes
          0 : Standard
          1 : Standard (with alternative initiation codons)
          2 : Vertebrate Mitochondrial
          3 : Yeast Mitochondrial
          4 : Mold, Protozoan, Coelenterate Mitochondrial and Mycoplasma/
              Spiroplasma
          5 : Invertebrate Mitochondrial
          6 : Ciliate Macronuclear and Dasycladacean
          9 : Echinoderm Mitochondrial
         10 : Euplotid Nuclear
         11 : Bacterial
         12 : Alternative Yeast Nuclear
         13 : Ascidian Mitochondrial
         14 : Flatworm Mitochondrial
         15 : Blepharisma Macronuclear
         16 : Chlorophycean Mitochondrial
         21 : Trematode Mitochondrial
         22 : Scenedesmus obliquus
         23 : Thraustochytrium Mitochondrial
Code to use [0]:
Search pattern: RA
Number of mismatches [0]:
Output file [rnops.fuzztran]:
```

Mandatory qualifiers:

[-sequence] (seqall)
: Sequence database USA.

-pattern (string)
: The standard IUPAC one-letter codes for the amino acids are used. The letter `x` is used for a position where any amino acid is accepted. Ambiguities are indicated by listing the acceptable amino acids for a given position between square brackets. For example, `[ALT]` stands for Ala, Leu, or Thr. Ambiguities are also indicated by listing the amino acids that are not accepted at a given position in curly brackets. For example, `{AM}` stands for any amino acid except Ala and Met. Each element in a pattern is separated from its neighbor by a dash. (Optional in *fuzztran*.)

  Repetition of an element of the pattern is indicated by following that element with a numerical value or a numerical range between parenthesis. For example, `x(3)` corresponds to `x-x-x`, while `x(2,4)` corresponds to `x-x`, `x-x-x`, or `x-x-x-x`. When a pattern is restricted to either the N- or C-terminal of a sequence, that pattern either starts with a "<" symbol or ends with a ">" symbol, respectively. A period ends the pattern. (Optional in *fuzztran*.) For example, `[DE](2)HS{P}X(2)PX(2,4)C8`.

-mismatch (integer)
: Number of mismatches.

[-outfile] (report)
: Output report filename.

Optional qualifiers:

-frame (menu)
Frame(s) to translate.

-table (menu)
Code to use.

garnier

garnier is an implementation of the original Garnier-Osguthorpe-Robson algorithm (GOR I) for predicting protein secondary structure.

Here is a sample session with *garnier*:

```
% garnier
Input sequence: sw:amic_pseae
Output file [amic_pseae.garnier]:
```

Mandatory qualifiers:

[-sequencea] (seqall)
Sequence database USA.

[-outfile] (report)
Output report filename.

Advanced qualifiers:

-idc (integer)
Set an idc parameter.

geecee

geecee calculates the fraction of G+C bases of the input nucleic acid sequence(s).

Here is a sample session with *geecee*:

```
% geecee embl:hhtetra
Output file [hhtetra.geecee]:
```

Mandatory qualifiers:

[-sequence] (seqall)
Sequence database USA.

[-outfile] (outfile)
Output filename.

getorf

getorf finds and outputs the sequences of open reading frames (ORFs).

Here is a sample session with *getorf*:

```
% getorf -minsize 300
Input sequence: embl:eclaci
Output sequence [eclaci.orf]:
```

Mandatory qualifiers:

[-sequence] (seqall)
Sequence database USA.

[-outseq] (seqoutall)
Output sequence(s) USA.

Optional qualifiers:

-table (menu)
Code to use. See the *fuzztran* description for codes.

-minsize (integer)
Minimum nucleotide size of ORF to report.

-find (menu)
This is a small menu of possible output options. The first four options are to select either the protein translation or the original nucleic acid sequence of the open reading frame. There are two possible definitions of an open reading frame: it may be a region that is free of STOP codons or it may be a region that begins with a START codon and ends with a STOP codon. The last three options are probably only of interest to those who want to investigate the statistical properties of the regions around potential START or STOP codons. The last option assumes that ORF lengths are calculated between two STOP codons.

Advanced qualifiers:

-[no]methionine (boolean)
START codons at the beginning of protein products will usually code for Methionine, despite what the codon will code for when it is internal to a protein. This qualifier sets all such START codons to code for Methionine by default.

-circular (boolean)
Is the sequence circular?

-[no]reverse (boolean)
Set this to be false if you do not want to find ORFs in the reverse complement of the sequence.

-flanking (integer)
If you chose an option in the type of sequence to find that provides the flanking sequence around a STOP or START codon, this allows you to set the number of nucleotides on either side of that codon to output. If the region of flanking nucleotides crosses the start or end of the sequence, no output is given for this codon.

helixturnhelix

helixturnhelix uses the method of Dodd and Egan to find helix-turn-helix nucleic acid binding motifs in proteins.

Here is a sample session with *helixturnhelix*:

```
% helixturnhelix
Input sequence: sw:laci_ecoli
Output file [laci_ecoli.hth]:
```

Mandatory qualifiers:

[-sequence] (seqall)
Sequence database USA.

[-outfile] (report)
Output report filename.

Optional qualifiers:

-mean (float)
Mean value.

-sd (float)
Standard Deviation (SD) value.

-minsd (float)
Minimum SD.

-eightyseven (boolean)
Use the old (1987) weight data.

hetparse

hteparse converts raw dictionary of heterogen groups to a file in EMBL-like format.

This is part of Jon Ison's protein structure analysis package. This package is still being developed. Please ignore this program until further details can be documented. All further queries should go to Jon Ison (*jison@hgmp.mrc.ac.uk*).

Here is a sample session with *hetparse*:

```
% hetparse
```

Mandatory qualifiers (bold if not always prompted):

[-inf] (infile)
Name of input file (raw dictionary of heterogen groups).

-dogrep (boolean)
Search a directory of files with keywords.

-path *(string)*
Directory to search with keywords.

-extn *(string)*
Exension of files to search in directory.

[-outf] (outfile)
Name of output file (EMBL format).

hmmgen

hmmgen generates a hidden Markov model for each alignment in a directory.

Here is a sample session with *hmmgen*:

```
% hmmgen
```

Mandatory qualifiers:

-infpath (string)
Location of sequence alignment files (input).

-infextn (string)
Extension of sequence alignment files.

-outfpath (string)
: Location of HMM files (output).

-outfextn (string)
: Extension of HMM files.

hmoment

hmoment plots or writes out the hydrophobic moment. Hydrophobic moment is the hydrophobicity of a peptide measured for a specified angle of rotation per residue.

Here is a sample session with *hmoment*:

```
% hmoment sw:hbb_human
Hydrophobic moment calculation
Output file [hbb_human.hmoment]:
```

Mandatory qualifiers (bold if not always prompted):

[-seqall] (seqall)
: Sequence database USA.

-graph *(xygraph)*
: Graph type.

-outfile *(outfile)*
: Output filename.

Optional qualifiers:

-window (integer)
: Window.

-bangle (integer)
: Beta sheet angle (degrees).

Advanced qualifiers:

-aangle (integer)
: Alpha helix angle (degrees).

-baseline (float)
: Graph marker line.

-plot (boolean)
: Produce graphic.

-double (boolean)
: Plot two graphs.

iep

iep calculates the isoelectric point of a protein from its amino acid composition assuming that no electrostatic interactions change the propensity for ionization.

Here is a sample session with *iep*:

```
% iep sw:laci_ecoli
Output file [laci_ecoli.iep]:
```

Mandatory qualifiers (bold if not always prompted):

[-sequencea] (seqall)
Sequence database USA.

-graph *(xygraph)*
Graph type.

-outfile *(outfile)*
Output filename.

Advanced qualifiers:

-step (float)
pH step value.

-amino (integer)
Number of N-termini.

-[no]termini (boolean)
Include charge at N and C terminus.

-plot (boolean)
Plot charge vs pH.

-[no]report (boolean)
Write results to a file.

infoalign

infoalign is small utility to list some simple properties of sequences in an alignment.

Here is a sample session with *infoalign*:

```
% infoalign globin.seq
```

Don't display the USA of a sequence:

```
% infoalign globin.seq -nousa -out stdout
```

Display only the name and sequence length of a sequence:

```
% infoalign globin.seq -only -name -seqlength -out stdout
```

Display only the name, number of gap characters and differences to the consensus sequence:

```
% infoalign globin.seq -only -name -gapcount -diffcount -out stdout
```

Display the name and number of gaps within a sequence:

```
% infoalign globin.seq -only -name -gaps -out stdout
```

Display information formatted with HTML:

```
% infoalign globin.seq -html -out stdout
```

Use the first sequence as the reference sequence to compare to:

```
% infoalign globin.seq -refseq 1 -out stdout
```

Mandatory qualifiers:

[-sequence] (seqset)
The sequence alignment to display.

[-outfile] (outfile)
If you enter the name of a file here, this program will write the sequence details into the specified file.

Optional qualifiers:

-refseq (string)
: If you give the number in the alignment or the name of a sequence, it will be taken to be the reference sequence. The reference sequence is the one to which all the other sequences are compared. If this is set to `0`, the consensus sequence is used as the reference sequence. By default, the consensus sequence is used as the reference sequence.

-matrix (matrix)
: This is the scoring matrix file used when comparing sequences. By default, it is the file *EBLOSUM62* (for proteins) or the file *EDNAFULL* (for nucleic sequences). These files are found in the *data* directory of the EMBOSS installation.

-html (boolean)
: Format output as an HTML table.

Advanced qualifiers:

-plurality (float)
: Set a cutoff for the percentage of positive scoring matches below which there is no consensus. The default plurality is taken as 50% of the total weight of all sequences in the alignment.

-identity (float)
: Sets the number of identities required at a position in order to return a consensus. If this is set to 100%, only columns of identities contribute to the consensus.

-only (boolean)
: This is a way of shortening the command line if you want only a few things to be displayed. Instead of using the options `-nohead -nousa -noname -noalign -nogaps -nogapcount -nosimcount -noidcount -nodiffcount` to get only the sequence length output, you can specify `-only -seqlength`.

-heading (boolean)
: Display column headings.

-usa (boolean)
: Display the USA of the sequence.

-name (boolean)
: Display `name` column.

-seqlength (boolean)
: Display `seqlength` column.

-alignlength (boolean)
: Display `alignlength` column.

-gaps (boolean)
: Display number of gaps.

-gapcount (boolean)
: Display number of gap positions.

-idcount (boolean)
: Display number of identical positions.

-simcount (boolean)
: Display number of similar positions.

-diffcount (boolean)
: Display number of different positions.

-change (boolean)
Display percentage of changed positions.

-description (boolean)
Display `description` column.

infoseq

infoseq is a small utility that lists the USA, name, accession number, type (nucleic or protein), length, percentage C+G, and/or description of a sequence.

Display information on a sequence:

```
% infoseq embl:paamir
```

Don't display the USA of a sequence:

```
% infoseq embl:paamir -nousa
```

Display only the name and length of a sequence:

```
% infoseq embl:paamir -only -name -length
```

Display only the description of a sequence:

```
% infoseq embl:paamir -only -desc
```

Display the type of a sequence:

```
% infoseq embl:paamir -only -type
```

Display information formatted with HTML:

```
% infoseq embl:paamir -html
```

Mandatory qualifiers:

[-sequence] (seqall)
Sequence database USA.

Optional qualifiers:

-outfile (outfile)
If you enter the name of a file here, this program will write the sequence details into that file.

-html (boolean)
Format output as an HTML table.

Advanced qualifiers:

-only (boolean)
This is a way of shortening the command line if you only want a few things to be displayed. Instead of specifying `-nohead -noname -noacc -notype -nopgc -nodesc` to get only the length output, you can specify `-only -length`.

-heading (boolean)
Display column headings.

-usa (boolean)
Display the USA of the sequence.

-name (boolean)
Display `name` column.

-accession (boolean)
Display `accession` column.

-gi (boolean)
Display GI column.

-version (boolean)
Display version column.

-type (boolean)
Display type column.

-length (boolean)
Display length column.

-pgc (boolean)
Display percent GC content column.

-description (boolean)
Display description column.

interface

interface reads coordinate files and writes files of inter-chain residue-residue contact data.

Here is a sample session with *interface*:

```
% interface
```

Mandatory qualifiers:

[-in] (infile)
Name of protein coordinate file for input (EMBL format).

-vdwf (string)
Name of data file with Van der Waals radii.

-thresh (float)
Threshold contact distance.

[-out] (outfile)
Name of contact file for output.

-conerrf (outfile)
Name of log file for the build.

Optional qualifiers:

-ignore (float)
If any two atoms from two different residues are at least this distance apart, no futher inter-atomic contacts are checked for for that residue pair. This speeds up the calculation considerably.

isochore

isochore plots GC content over a sequence. It is intended for large sequences such as complete chromosomes or large genomic contigs, although interesting results can also be obtained from shorter sequences.

Here is a sample session with *isochore*. This reads yeast chromosome 3 from a remote web server, so the sequence reading may take a little time:

```
% isochore
Input sequence: tgb:scchriii
Output file [scchriii.iso]:
click here for result
```

Mandatory qualifiers:

[-sequence] (sequence)
: Sequence USA.

[-out] (outfile)
: Output filename.

-graph (xygraph)
: Graph type.

Optional qualifiers:

-window (integer)
: Window size.

-shift (integer)
: Shift increment.

---

lindna

lindna draws linear maps of DNA constructs.

Here is a sample session with *lindna*:

```
% lindna
Draws linear maps of DNA constructs
Graph type [x11]:
Input file [inputfile]: data.linp
do you want a ruler (Y or N) [Y]:
type of blocks (enter Open, Filled, or Outline) [Filled]:
```

Mandatory qualifiers:

-graphout (graph)
: Graph type.

-inputfile (infile)
: Input file containing mapping data.

-ruler (string)
: Do you want a ruler (`Y` or `N`)?

-blocktype (string)
: Type of blocks: `Open`, `Filled`, or `Outline`. Option `Outline` draws filled blocks surrounded by a black border.

Optional qualifiers:

-intersymbol (string)
: Type of junctions between blocks (one of `Straight`, `Up`, `Down`, or `No`).

-intercolor (integer)
: Color of junctions between blocks (enter a color number).

-interticks (string)
: Do you want horizontal junctions between ticks (`Y` or `N`)?

-gapsize (integer)
: Interval between ticks in the ruler (enter an integer).

-ticklines (string)
: Do you want vertical lines at the ruler's ticks (Y or N)?

-textheight (float)
: Height of text. Enter a number less than 1 to decrease the size. Enter a number greater than 1 to increase the size.

-textlength (float)
: Length of text. Enter a number less than 1 to decrease the size. Enter a number greater than 1 to increase the size.

-margin (float)
: Width of left margin. This is the region left of the groups where the names of the groups are displayed. Enter a number less than 1 to decrease the size. Enter a number greater than 1 to increase the size.

-tickheight (float)
: Height of ticks. Enter a number less than 1 to decrease the size. Enter a number greater than 1 to increase the size.

-blockheight (float)
: Height of blocks. Enter a number less than 1 to decrease the size. Enter a number greater than 1 to increase the size.

-rangeheight (float)
: Height of range ends. Enter a number less than 1 to decrease the size. Enter a number greater than 1 to increase the size.

-gapgroup (float)
: Space between groups. Enter a number less than 1 to decrease the size. Enter a number greater than 1 to increase the size.

-postext (float)
: Space between text and ticks, blocks, and ranges. Enter a number less than 1 to decrease the size. Enter a number greater than 1 to increase the size.

---

listor

listor reads in two sets of sequences and writes out a list file (file of filenames) that result from the logical union of these two sets of sequences. It is a simple way of manipulating and editing lists or sets of sequences to produce a list file.

Here is a sample session with *listor*:

```
% listor file1 file1
Writes a list file of the logical OR of two sets of sequences
Output file [outfile.list]:
```

Mandatory qualifiers:

[-firstset] (seqset)
: Sequence set USA.

[-secondset] (seqset)
: Sequence set USA.

[-outlist] (outfile)
: The list of sequence names is written to this list file.

Optional qualifiers:

-operator (menu)

The following logical operators combine the sequences in the following ways:

OR gives all that occur in one set or the other.

AND gives only those which occur in both sets.

XOR gives those which only occur in one set or the other, but not in both.

NOT gives those which occur in the first set except for those that also occur in the second.

marscan

marscan finds MAR/SAR sites in nucleic sequences.

Here is a sample session with *marscan*:

```
% marscan
marscan
Finds MAR/SAR sites in nucleic sequences
Input sequence(s): EMBL:HSHBB
Output file [hshbb.marscan]:
```

Mandatory qualifiers:

[-sequence] (seqall)

Sequence database USA.

[-outfile] (report)

File for output of MAR/SAR recognition signature (MRS) regions. This contains details of the MRS in normal GFF format. The MRS consists of two recognition sites: one of 8 bp and one of 16 bp on either sense strand of the genomic DNA, within 200 bp of one another.

maskfeat

maskfeat masks features of a sequence.

Here is a sample session with *maskfeat*. It masks out a feature whose type is "repeat_region" from position 2331 to 2356:

```
% maskfeat em:AB000360
Mask off features of a sequence.
Output sequence [ab000360.fasta]:
```

Mandatory qualifiers:

[-sequence] (seqall)

Sequence database USA.

[-outseq] (seqout)

Output sequence USA.

Optional qualifiers:

-type (string)

By default, any feature in the feature table with a type starting repeat is masked. You can set this to be any feature type you want to mask. See Chapter 2 for a list of the EMBL feature types, and Chapter 3 for a list of the SWISS-PROT feature types. The type may be wildcarded by using *. If you want to mask more than one type, separate their names with spaces or commas. For example:

```
*UTR repeat*
```

-maskchar (string)
Character to use when masking. Default is X for protein sequences, and N for nucleic sequences.

maskseq

maskseq allows you to mask off regions of a sequence with a specified letter.

To mask off bases 10 to 12 from a sequence *gatta.seq* and write to the new sequence file *gatta2.seq*:

```
% maskseq gatta.seq gatta2.seq -reg=10-12
```

To mask off bases 20 to 30 from a sequence *hdh.seq* using the character n and write to the new sequence file *hdh2.seq*:

```
% maskseq hdh.seq hdh2.seq -reg=20-30 -mask=n
```

To mask off the regions 20 to 23, 34 to 45 and 88 to 90 in *yuy.seq*:

```
% maskseq yuy.seq yuy2.seq -reg=20-23,34-45,88-90
```

Mandatory qualifiers:

[-sequence] (sequence)
Sequence USA.

-regions (range)
Regions to mask. A set of regions is specified by a set of pairs of positions. The positions are integers. They are separated by any non-digit, non-alpha character. Examples of region specifications are: 24-45, 56-78, 1:45, 67=99, 765..888, and 1,5,8,10,23,45,57,99.

[-outseq] (seqout)
Output sequence USA.

Optional qualifiers:

-maskchar (string)
Character to use when masking. Default is X for protein sequences, and N for nucleic sequences.

matcher

matcher compares two sequences looking for local sequence similarities, using a rigorous algorithm.

Here is a sample session with *matcher*:

```
% matcher tsw:hba_human tsw:hbb_human
Finds the best local alignments between two sequences
Output file [hba_human.matcher]:
```

Here is an example to find the 10 best alignments:

```
% matcher tsw:hba_human tsw:hbb_human -alt 10
Finds the best local alignments between two sequences
Output file [hba_human.matcher]: hba_human.matcher10
```

Mandatory qualifiers:

[-sequencea] (sequence)
: Sequence USA.

[-sequenceb] (sequence)
: Sequence USA.

[-outfile] (align)
: Output alignment filename.

Optional qualifiers:

-datafile (matrix)
: This is the scoring matrix file used when comparing sequences. By default, it is the file *EBLOSUM62* (for proteins) or the file *EDNAFULL* (for nucleic sequences). These files are found in the *data* directory of the EMBOSS installation.

-alternatives (integer)
: This sets the number of alternative matches output. By default, only the highest scoring alignment is shown. A value of `2` gives you other reasonable alignments. In some cases (for example, multidomain proteins of cDNA and gemomic DNA comparisons) other interesting and significant alignments may be found.

-gappenalty (integer)
: The gap penalty is the score taken away when a gap is created. The best value depends on the choice of comparison matrix. The default value of `14` assumes you are using the EBLOSUM62 matrix for protein sequences, or a value of `16` and the EDNAFULL matrix for nucleotide sequences.

-gaplength (integer)
: The gap length, or gap extension penalty is added to the standard gap penalty for each base or residue in the gap. This is how long gaps are penalized. You can usually expect a few long gaps rather than many short gaps, so the gap extension penalty should be lower than the gap penalty. An exception occurs when one or both sequences are single reads with possible sequencing errors, in which case you would expect many single base gaps. You can obtain this result by setting the gap penalty to zero (or a very low value) and using the gap extension penalty to control gap scoring.

megamerger

megamerger takes two overlapping sequences and merges them into one sequence. It could thus be regarded as the opposite of what *splitter* does.

Here is a sample session with *megamerger*. There are many mismatches between these two sequences and the merged sequence should therefore be treated with great caution:

```
% megamerger embl:ap000504 embl:af129756
Merge two large overlapping nucleic acid sequences
Word size [20]:
Output sequence [ap000504.merged]:
Output file [ap000504.megamerger]:
```

Mandatory qualifiers:

[-seqa] (sequence)
  Sequence USA.

[-seqb] (sequence)
  Sequence USA.

-wordsize (integer)
  Word size.

[-outseq] (seqout)
  Output sequence USA.

[-report] (outfile)
  Output report.

merger

merger joins two overlapping nucleic acid sequences into one merged sequence.

Here is a sample session with *merger*:

```
% merger
Merge two overlapping nucleic acid sequences
Input sequence: tembl:eclacy
Second sequence: tembl:eclaca
Output sequence [eclacy.fasta]:
Output alignment [eclacy.out2]:
```

Typically, one of the sequences will need to be reverse-complemented to put it into the correct orientation to make it join. For example:

```
% merger file1.seq file2.seq -sreverse2 -outseq merged.seq -outfile stdout
```

Mandatory qualifiers:

[-seqa] (sequence)
  Sequence USA.

[-seqb] (sequence)
  Sequence USA.

[-outseq] (seqout)
  Output sequence USA.

[-outfile] (align)
  Output alignment and explanation.

Optional qualifiers:

-datafile (matrixf)
  This is the scoring matrix file used when comparing sequences. By default, it is the file *EBLOSUM62* (for proteins), or the file *EDNAFULL* (for nucleic sequences). These files are found in the *data* directory of the EMBOSS installation.

-gapopen (float)
  Gap opening penalty.

-gapextend (float)
  Gap extension penalty.

msbar

msbar changes a sequence a lot or a little, attempting to emulate various forms of mutation. You can set the number and types of mutations.

Here is a sample session with *msbar*. This asks for 5 mutations, with point mutations as changes (substitutions) and the codon and block mutations ignored:

```
% msbar
Input sequence: embl:eclaci
Output sequence [eclaci.fasta]:
Number of times to perform the mutation operations [1]: 5
Point mutation operations
          0 : None
          1 : Any of the following
          2 : Insertions
          3 : Deletions
          4 : Changes
          5 : Duplications
          6 : Moves
Types of point mutations to perform [0]: 4
Codon mutation operations
          0 : None
          1 : Any of the following
          2 : Insertions
          3 : Deletions
          4 : Changes
          5 : Duplications
          6 : Moves
Types of codon mutations to perform [0]:
Block mutation operations
          0 : None
          1 : Any of the following
          2 : Insertions
          3 : Deletions
          4 : Changes
          5 : Duplications
          6 : Moves
Types of block mutations to perform [0]:
```

Mandatory qualifiers (bold if not always prompted):

[-sequence] (seqall)
: Sequence database USA.

-count (integer)
: Number of times to perform the mutation operations.

-point (menu)
: Types of point mutations to perform.

-block (menu)
: Types of block mutations to perform.

-codon *(menu)*
: Types of codon mutations to perform. These are only done if the sequence is nucleic.

[-outseq] (seqoutall)
: Output sequence(s) USA.

Optional qualifiers (bold if not always prompted):

-inframe *(boolean)*
: Do `codon` and `block` operations in frame.

Advanced qualifiers:

-minimum (integer)
: Minimum size for a block mutation.

-maximum (integer)
: Maximum size for a block mutation.

EMBOSS

---

mwcontam

mwcontam finds molecular weights that are common between a set of mass spectrometry result files.

Here is a sample session with *mwcontam*:

```
% mwcontam
Shows molwts that match across a set of files
Comma separated file list: mw1.dat,mw2.dat,mw3.dat
ppm tolerance [50.0]:
Output file [outfile.mwcontam]:
```

Mandatory qualifiers:

[-files] (filelist)
: Comma-separated file list.

-tolerance (float)
: ppm tolerance.

[-outfile] (outfile)
: Output filename.

---

mwfilter

mwfilter is designed to remove unwanted (noisy) data from mass spectrometry output in proteomics.

Here is a sample session with *mwfilter*:

```
% mwfilter
Filter noisy molwts from mass spec output
Input file: molwts.dat
ppm tolerance [50.0]:
Output file [molwts.mwfilter]:
```

Mandatory qualifiers:

[-infile] (infile)
: Molecular weight file input.

-tolerance (float)
: ppm tolerance.

[-outfile] (outfile)
: Output filename.

Optional qualifiers:

-showdel (boolean)
: Output deleted molecular weights.

Advanced qualifiers:

-datafile (string)
: Data file of noisy molecular weights.

needle

needle uses the Needleman-Wunsch global alignment algorithm to find the optimum alignment (including gaps) of two sequences when considering their entire length.

Here is a sample session with *needle*:

```
% needle sw:hba_human sw:hbb_human

Gap opening penalty [10.0]:
Gap extension penalty [0.5]:
Output file [hba_human.needle]:
```

Mandatory qualifiers:

[-sequencea] (sequence)
: Sequence USA.

[-seqall] (seqall)
: Sequence database USA.

-gapopen (float)
: The gap open penalty is the score taken away when a gap is created. The best value depends on the choice of comparison matrix. The default value assumes you are using the EBLOSUM62 matrix for protein sequences, and the EDNAFULL matrix for nucleotide sequences.

-gapextend (float)
: The gap extension penalty is added to the standard gap penalty for each base or residue in the gap. This is how long gaps are penalized. You can usually expect a few long gaps rather than many short gaps, so the gap extension penalty should be lower than the gap penalty. An exception occurs when one or both sequences are single reads with possible sequencing errors, in which case you should expect many single base gaps. You can obtain this result by setting the gap open penalty to zero (or a very low value) and using the gap extension penalty to control gap scoring.

[-outfile] (align)
: Output alignment filename.

Optional qualifiers:

-datafile (matrixf)
: This is the scoring matrix file used when comparing sequences. By default, it is the file *EBLOSUM62* (for proteins) or the file *EDNAFULL* (for nucleic sequences). These files are found in the *data* directory of the EMBOSS installation.

Advanced qualifiers:

-[no]similarity (boolean)
: Display percent identity and similarity.

---

newcpgreport

newcpgreport is used in the production of the CpG Island database CPGISLE. It produces CPGISLE database entry format reports for a potential CpG island. See the FTP site: *ftp://ftp.ebi.ac.uk/pub/databases/cpgisle/* for the finished database.

Here is a sample session with *newcpgreport*:

```
% newcpgreport
Input sequence: embl:rnu68037
Window size [100]:
Shift increment [1]:
Minimum Length [200]:
Minimum observed/expected [0.6]:
Minimum percentage [50.]:
Output file [rnu68037.newcpgreport]:
```

Mandatory qualifiers:

[-sequence] (seqall)
: Sequence database USA.

-window (integer)
: Window size.

-shift (integer)
: Shift increment.

-minlen (integer)
: Minimum length.

-minoe (float)
: Minimum observed or expected.

-minpc (float)
: Minimum percentage.

[-outfile] (outfile)
: Output filename.

---

newcpgseek

newcpgseek reports CpG rich regions of a sequence as candidate CpG islands.

Here is a sample session with *newcpgseek*:

```
% newcpgseek
Input sequence: embl:rnu68037
CpG score [17]:
Output file [rnu68037.newcpgseek]:
```

Mandatory qualifiers:

[-sequence] (seqall)
: Sequence database USA.

-score (integer)
: CpG score.

[-outfile] (outfile)
: Output filename.

---

newseq

newseq allows you to type a sequence into a file in a quick and easy manner.

Type in a short sequence to the file *mycc.pep* in EMBL format:

```
% newseq
Type in a short new sequence.
Output sequence [outfile.fasta]: embl::mycc.pep
Name of the sequence: cytoc
Description of the sequence: fragment of cytochrome C
Type of sequence

         N : Nucleic
         P : Protein
Type of sequence [N]: p
Enter the sequence: KKKEERADLIAY
```

Display the resulting new file:

```
% more mycc.pep
ID   cytoc           STANDARD;      PRT;    12 AA.
DE   fragment of cytochrome C
SQ   SEQUENCE     12 AA;    1464 MW;  2BF1DB53 CRC32;
     KKKEERADLI AY
//
```

Mandatory qualifiers:

[-outseq] (seqout)
: Output sequence USA.

[-name] (string)
: The name of of the sequence should be a single word that you will use to identify the sequence. It should have no (or few) punctuation characters in it.

[-description] (string)
: Enter any description of the sequence that you require.

[-type] (menu)
: Type of sequence.

[-sequence] (string)
: The sequence itself. Because of the limitation of the operating system, you will only be able to type in a short sequence of around 250 characters. The keyboard will beep at you when you have reached this limit and you will not be able to press the Return or Enter key until you have deleted a few characters.

noreturn

noreturn removes carriage return from ASCII files.

Here is a sample session with *noreturn*:

```
% noreturn
Input file: abc.dat
Output file [abc.noreturn]:
```

Mandatory qualifiers:

[-infile] (infile)
: Infile value (no help text).

[-outfile] (outfile)
: Output filename.

EMBOSS

notseq

notseq splits the input sequences into those that you want to keep and those you want to exclude.

Here is a sample session with *notseq*. In this case the excluded sequences (*clone186* and *clone876*) are simply thrown away and not saved to any file:

```
% notseq
Excludes a set of sequences and writes out the remaining ones
Input sequence(s): globins.fasta
Sequence names to exclude: myg_phyca,lgb2_luplu
Output sequence [hbb_human.fasta]: mydata.seq
```

Here is an example where the sequences to exclude are saved to another file:

```
% notseq -junkout excluded.seq
Excludes a set of sequences and writes out the remaining ones
Input sequence(s): globins.fasta
Sequence names to exclude: hb*
Output sequence [hbb_human.fasta]: mydata.seq
```

Mandatory qualifiers:

[-sequence] (seqall)
: Sequence database USA.

[-exclude] (string)
: Enter a list of sequence names or accession numbers to exclude from the sequences read. The excluded sequences are written to the file specified in the `junkout` parameter. The remainder will be written out to the file specified in the `outseq` parameter. The list of sequence names can be separated by either spaces or commas. The sequence names can be wildcarded. The sequence names are case independent. An example of a list of sequences to be excluded is: `myseq, hs*, one two three`. A file containing a list of sequence names can be specified by giving the filename preceded by an at sign, e.g., `@names.dat`.

[-outseq] (seqoutall)
: Output sequence(s) USA.

Optional qualifiers:

-junkout (seqoutall)

This file collects the sequences that were excluded from the main output file of sequences.

nrscope

nrscope reads in the EMBL-like format SCOP classification file generated by the EMBOSS application *scope*, and writes a file of nonredundant domains in the same format. Domain sequences are extracted from the clean domain coordinate files generated by the EMBOSS application *domainer*.

Here is a sample session with *nrscope*:

```
% nrscope
Converts redundant EMBL-format SCOP file to non-redundant one
Name of scop file for input (embl-like format) [Escop.dat]: /data/scop/
Escop.dat
Name of non-redundant scop file for output (embl-like format) [EscopNR.dat]:
EscopNR.test
Location of clean domain coordinate files for input (embl-like format) [./]:
/data/cpdbscop/
File extension of clean domain coordinate files [.pxyz]:
The % sequence identity redundancy threshold [95]: 95
Residue substitution file [EBLOSUM62]:
Gap insertion penalty [10]: 20
Gap extension penalty [0.5]: 1
Name of log file for the build [nrscope.log]: EscopNR.log
D3SDHA_
D3SDHB_
D3HBIA_
D3HBIB_
D4SDHA_
D4SDHB_
D4HBIA_
D4HBIB_
D5HBIA_
D5HBIB_
```

Mandatory qualifiers:

[-scopin] (infile)

Name of scop file for input (EMBL-like format).

[-dpdb] (string)

Location of clean domain coordinate files for input (EMBL-like format).

[-extn] (string)

File extension of clean domain coordinate files.

[-thresh] (float)

The percentage sequence identity redundancy threshold.

[-datafile] (matrixf)

This is the scoring matrix file used when comparing sequences.

[-gapopen] (float)
: The gap insertion penalty is the score taken away when a gap is created. The best value depends on the choice of comparison matrix. The default value assumes you are using the EBLOSUM62 matrix for protein sequences, and the EDNAFULL matrix for nucleotide sequences.

[-gapextend] (float)
: The gap extension penalty is added to the standard gap penalty for each base or residue in the gap. This is how long gaps are penalized. You can usually expect a few long gaps rather than many short gaps, so the gap extension penalty should be lower than the gap penalty. An exception occurs when one or both sequences are single reads with possible sequencing errors, in which case you should expect many single base gaps. You can obtain this result by setting the gap open penalty to zero (or a very low value) and using the gap extension penalty to control gap scoring.

[-scopout] (outfile)
: Name of nonredundant scop file for output (EMBL-like format).

[-errf] (outfile)
: Name of log file for the build.

nthseq

nthseq extracts the indicated sequence from a multiple set of sequences and writes it out.

Here is a sample session with *nthseq*:

```
% nthseq
Input sequence: embl:eclac*
The number of the sequence to output [1]: 2
Output sequence [eclaca.fasta]:
```

Mandatory qualifiers:

[-sequence] (seqall)
: Sequence database USA.

-number (integer)
: The number of the sequence to output.

[-outseq] (seqout)
: Output sequence USA.

octanol

octanol displays protein hydropathy.

Here is a sample session with *octanol*:

```
% octanol
Input sequence: sw:opsd_human
Graph type [x11]:
```

Mandatory qualifiers:

[-sequencea] (sequence)
: Sequence USA.

[-graph] (xygraph)
: Graph type.

Optional qualifiers:

-datafile (datafile)
: White-Wimley data file (*Ewhite-wimley.dat*).

-width (integer)
: Window size.

-octanolplot (boolean)
: Display the octanol plot.

-interfaceplot (boolean)
: Display the interface plot.

-[no]differenceplot (boolean)
: Display the difference plot.

oddcomp

oddcomp searches a series of protein files, reporting the identifier for those that exceed a certain amino acid composition threshold in a portion of the sequence.

An example of the use of *oddcomp* to search for entries in SWISS-PROT containing at least 1 SR and at least 2 RS:

```
% oddcomp
Finds protein sequence regions with a biased composition
Input sequence(s): sw:*
Output file [5h1d_fugru.oddcomp]: out.odd
Input file: test.comp
Window size to consider (e.g. 30 aa) [30]:
```

Mandatory qualifiers:

[-sequence] (seqall)
: Sequence database USA.

[-compdata] (infile)
: This is a file in the format of the output produced by *compseq* used to set the minimum frequencies of words in this analysis.

[-window] (integer)
: This is the size of window in which to count. If you want to count frequencies in a 40 aa stretch, enter `40` here.

[-outfile] (outfile)
: This is the results file.

Advanced qualifiers:

-[no]ignorebz (boolean)
: The amino acid code B represents Asparagine or Aspartic acid, and the code Z represents Glutamine or Glutamic acid. These codes are not commonly used and you may not want to count words containing them. This command will note codes B and Z in the count of "Other" words.

palindrome

palindrome looks for inverted repeats (stem loops) in a nucleotide sequence.

Here is a sample session with *palindrome*. Because there are a number of overlapping possibilities in this sequence, we choose a longer minimum repeat length:

```
% palindrome
Input sequence: embl:hsts1
Enter minimum length of palindrome [10]: 15
Enter maximum length of palindrome [100]:
Enter maximum gap between repeated regions [100]:
Number of mismatches allowed [0]:
Output file [hsts1.pal]:
Report overlapping matches [Y]:
```

Mandatory qualifiers:

[-insequence] (seqall)
: Sequence database USA.

-minpallen (integer)
: Enter minimum length of palindrome.

-maxpallen (integer)
: Enter maximum length of palindrome.

-gaplimit (integer)
: Enter maximum gap between repeated regions.

-nummismatches (integer)
: Number of mismatches allowed.

[-outfile] (outfile)
: Output filename.

-[no]overlap (boolean)
: Report overlapping matches.

EMBOSS

pasteseq

pasteseq allows you to insert one sequence into another sequence after a specified position, then writes the results to a sequence file.

To insert the sequence *tst.seq* after position 67 in sequence *ese.seq* and write the results to the file *estste.seq*:

```
% pasteseq ese.seq tst.seq estste.seq -pos=67
```

This is probably slightly more readable if the argument names are used instead of relying on the parameter positions:

```
% pasteseq -seq=ese.seq -ins=tst.seq -out=estste.seq -pos=67
```

To insert the sequence *tst.seq* before the start of sequence *ese.seq*, use -pos=0:

```
% pasteseq -seq=ese.seq -ins=tst.seq -out=estste.seq -pos=0
```

Mandatory qualifiers:

[-sequence] (sequence)
: Sequence USA.

[-insseq] (sequence)
: Sequence to insert.

-pos (integer)
: The position in the main input sequence to insert after. To insert before the start, use the position 0.

[-outseq] (seqout)
: Output sequence USA.

patmatdb

patmatdb takes a protein motif and compares it to a set of protein sequences.

Here is a sample session with *patmatdb*:

```
% patmatdb
Search a protein sequence with a motif
Input sequence(s): sw:*
Protein motif to search for: st[ty]s
Output file [5h1d_fugru.patmatdb]:
```

Mandatory qualifiers:

[-sequence] (seqall)
: Sequence database USA

[-motif] (string)
: Patterns for *patmatdb* are based on the format of pattern used in the PROSITE database. For example: `[DE](2)HS{P}X(2)PX(2,4)C` means two Asps or Glus in any order followed by His, Ser, any residue other than Pro, then two of any residue followed by Pro followed by two to four of any residue followed by Cys. The search is case-independent, so `AAA` matches `aaa`.

[-outfile] (report)
: Output report filename.

patmatmotifs

patmatmotifs takes a protein sequence and compares it to the PROSITE database of motifs.

Here is a sample session with *patmatmotifs*:

```
% patmatmotifs -full
Matching Prosite Motif Database to a single sequence.
Input sequence: sw:12s1_arath
Output file [12s1_arath.patmatmotifs]:
```

Mandatory qualifiers:

[-sequence] (sequence)
: Sequence USA.

[-outfile] (report)
: Output report filename.

Optional qualifiers:

-full (boolean)
: Provide full documentation for matching patterns.

-[no]prune (boolean)
: Ignore simple patterns. If this is true, these simple post-translational modification sites are not reported: myristyl, asn_glycosylation, camp_phospho_site, pkc_phospho_site, ck2_phospho_site, and tyr_phospho_site.

pdbparse

pdbpars parses PDB files and writes cleaned-up protein coordinate files.

Here is a sample session with *pdbparse*:

```
% pdbparse
```

Mandatory qualifiers:

[-pdb] (string)
Location of PDB files (PDB format input).

[-pdbextn] (string)
Extension of PDB files (PDB format input).

-nocamask (boolean)
The group will not appear in either the CO or SQ records of the clean coordinate file.

-nocamaska (boolean)
The amino acid will not appear in the CO record, but will still be present in the SQ record of the clean coordinate file.

-atommask (boolean)
The amino acid will appear not appear in the CO record but will still be present in the SQ record of the clean coordinate file.

[-cpdb] (string)
Location of coordinate files (EMBL format output).

[-cpdbextn] (string)
Extension of coordinate files (EMBL format output).

-errf (outfile)
Name of log file for the build.

Optional qualifiers:

-chnsiz (integer)
Minimum number of residues in a chain for it to be parsed.

Advanced qualifiers:

-[no]cpdbnaming (boolean)
Use pdbid code to name files.

-maxmis (integer)
Maximum number of permissible mismatches between the `ATOM` and `SEQRES` sequences.

pdbtosp

pdbtosp converts raw SWISS-PROT:PDB equivalence file to EMBL-like format.

Here is a sample session with *pdbtosp*:

```
% pdbtosp
```

Mandatory qualifiers:

[-infilea] (infile)
Name of raw SWISS-PROT:PDB equivalence file (input).

[-outfile] (outfile)
Name of SWISS-PROT:PDB equivalence file (EMBL format output).

pepcoil

pepcoil calculates the probability of a coiled-coil structure for windows of 28 residues through a protein sequence using the method of Lupas A, et al. (1991).

Here is a sample session with *pepcoil*:

```
% pepcoil
Input sequence: sw:gcn4_yeast
Window size [28]:
Output file [gcn4_yeast.pepcoil]:
```

Mandatory qualifiers:

[-sequence] (seqall)
: Sequence database USA.

-window (integer)
: Window size.

[-outfile] (outfile)
: Output filename.

Advanced qualifiers:

-[no]coil (boolean)
: Report coiled coil regions.

-frame (boolean)
: Show coil frameshifts.

-[no]other (boolean)
: Report non coiled coil regions.

pepinfo

pepinfo detects and displays various useful metrics about a protein sequence.

Here is a sample session with *pepinfo*:

```
% pepinfo
Plots simple amino acid properties in parallel
Input sequence: sw:opsd_human
Graph type [x11]:
Output file [pepinfo.out]:
```

Mandatory qualifiers:

[-inseq] (sequence)
: Sequence USA.

-graph (xygraph)
: Graph type.

[-outfile] (outfile)
: Output filename.

Optional qualifiers:

-hwindow (integer)
: Window size for hydropathy averaging.

-aaproperties (string)
User-defined amino acid properties

-aahydropathy (string)
User-defined hydropathy data.

Advanced qualifiers:

-[no]generalplot (boolean)
Plot histogram of general properties.

-[no]hydropathyplot (boolean)
Plot graphs of hydropathy

pepnet

pepnet is a method of displaying the residues of a protein in a simple 3,4,3,4 repeating pattern that emulates at a simple level the arrangement of residues around an alpha helix.

Here is a sample session with *pepnet*:

```
% pepnet -sask
Input sequence: sw:gcn4_yeast
     Begin at position [start]: 253
       End at position [end]: 274
Graph type [x11]:
```

Mandatory qualifiers (bold if not always prompted):

[-sequence] (sequence)
Sequence USA.

-graph *(graph)*
Graph type.

Optional qualifiers (bold if not always prompted):

-squares *(string)*
By default, the aliphatic residues ILVM are marked with squares.

-diamonds *(string)*
By default, the residues DENQST are marked with diamonds.

-octags *(string)*
By default, the positively charged residues HKR are marked with octagons.

Advanced qualifiers:

-amphipathic (boolean)
If true, the residues ACFGILMVWY are marked as squares. All other residues are not marked. This overrides any other markup you may have specified using the -squares, -diamonds, or -octags qualifiers.

-data (boolean)
Output data to a file instead of plotting it.

pepstats

pepstats outputs a report of simple protein sequence information such as molecular weight, number of residues, average residue weight, charge, isoelectric point, etc.

Here is a sample session with *pepstats*:

```
% pepstats
Protein statistics
Input sequence: sw:laci_ecoli
Output file [laci_ecoli.pepstats]:
```

Mandatory qualifiers:

[-sequencea] (sequence)
Sequence USA.

-outfile (outfile)
Output filename.

Advanced qualifiers:

-[no]termini (boolean)
Include charge at N and C termini.

-aadata (string)
Molecular weight data for amino acids.

pepwheel

pepwheel displays peptide sequences in a helical representation.

Here is a sample session with *pepwheel*:

```
% pepwheel sw:hbb_human -send 30
Shows protein sequences as helices
Graph type [x11]:
```

Mandatory qualifiers (bold if not always prompted):

[-sequence] (sequence)
Sequence USA.

-outfile *(outfile)*
Output filename.

-graph *(graph)*
Graph type.

Optional qualifiers (bold if not always prompted):

-steps (integer)
The number of residues plotted per turn is this value divided by the `turns` value.

-turns (integer)
The number of residues plotted per turn is the `steps` value divided by this value.

-squares *(string)*
By default, the aliphatic residues ILVM are marked with squares.

-diamonds *(string)*
By default, the residues DENQST are marked with diamonds.

-octags *(string)*
By default, the positively charged residues HKR are marked with octagons.

Advanced qualifiers:

-[no]wheel (boolean)
Plot the wheel.

-amphipathic (boolean)
If this is true, the residues ACFGILMVWY are marked as squares, and all other residues are unmarked. This overrides any other markup you may have specified using the `-squares`, `-diamonds`, or `-octags` qualifiers.

-data (boolean)
Output the match data to a file instead of plotting it.

pepwindow

pepwindow reads in a protein sequence and displays a graph of the classic Kyte & Doolittle hydropathy plot of that protein.

Here is a sample session with *pepwindow*:

```
% pepwindow sw:hba_human
Displays protein hydropathy
Graph type [x11]:
```

Mandatory qualifiers:

[-sequencea] (sequence)
Sequence USA.

-graph (xygraph)
Graph type.

Optional qualifiers:

-datafile (datafile)
Nakai database file.

-length (integer)
Window size.

pepwindowall

pepwindowall produces a set of superimposed Kyte & Doolittle hydropathy plots from an aligned set of protein sequences.

Here is a sample session with *pepwindowall*:

```
% pepwindowall globin.msf -gxtitle="Base Number" -gytitle="hydropathy"
Displays protein hydropathy of a set of sequences
Graph type [x11]:
```

Mandatory qualifiers:

[-msf] (seqset)
File containing a sequence alignment.

[-graph] (xygraph)
Graph type.

Optional qualifiers:

-datafile (datafile)
Nakai database file.

-length (integer)
Window size.

plotcon

plotcon plots the quality of conservation of a sequence alignment.

Here is a sample session with *plotcon*:

```
% plotcon -sformat msf alignment.msf
```

Mandatory qualifiers (bold if not always prompted):

[-msf] (seqset)
File containing a sequence alignment.

-winsize (integer)
Number of columns to average alignment quality over. The larger this value is, the smoother the plot will be.

-graph *(xygraph)*
Graph type.

-outfile *(outfile)*
Display as data.

Optional qualifiers:

-scorefile (matrix)
This is the scoring matrix file used when comparing sequences. By default, it is the file *EBLOSUM62* (for proteins) or the file *EDNAFULL* (for nucleic sequences). These files are found in the *data* directory of the EMBOSS installation.

Advanced qualifiers:

-data (boolean)
Output the match data to a file, instead of plotting it.

plotorf

plotorf plots potential open reading frames.

Here is a sample session with *plotorf*:

```
% plotorf
Plot potential open reading frames
Input sequence: embl:paamir
Graph type [x11]:
```

Specifying your own START and STOP codons with a mitochondrial sequence:

```
% plotorf -start ATT,ATC,ATA,ATG,GTG -stop TAA,TAG,AGA,AGG
Plot potential open reading frames
Input sequence: mito.seq
Graph type [x11]:
```

Mandatory qualifiers:

[-sequence] (sequence)
Sequence USA.

-graph (xygraph)
Graph type.

Advanced qualifiers:

-start (string)
Start codons.

-stop (string)
Stop codons.

EMBOSS

polydot

polydot compares all sequences in a set of sequences. It then draws a dotplot for each pair of sequences by marking where words (tuples) of a specified length have an exact match in both sequences and optionally reports all identical matches to feature files.

Here is a sample session with *polydot*:

```
% polydot globin.fasta -gtitle="Polydot of globin.fasta"
```

Mandatory qualifiers (bold if not always prompted):

[-sequences] (seqset)
File containing a sequence alignment.

-wordsize (integer)
Word size.

-graph *(graph)*
Graph type.

-outfile *(outfile)*
Output filename.

Optional qualifiers:

-[no]boxit (boolean)
Draw a box around each dotplot.

-dumpfeat (boolean)
Dump all matches as feature files.

-format (string)
Format in which to dump.

-ext (string)
Extension for feature file.

Advanced qualifiers:

-data (boolean)
: Output the match data to a file instead of plotting it.

-gap (integer)
: This specifies the size of the gap that is used to separate the individual dotplots in the display. The size is measured in residues, as displayed in the output.

preg

preg searches for matches of a regular expression to a protein sequence.

Here is a sample session with *preg*:

```
% preg
regular expression search of a protein sequence
Input sequence(s): sw:*_rat
Output file [100k_rat.preg]: stdout
Regular expression pattern: IA[QWF]A
```

Mandatory qualifiers:

[-sequence] (seqall)
: Sequence database USA.

[-pattern] (regexp)
: Regular expression pattern.

[-outfile] (outfile)
: Output filename.

prettyplot

prettyplot reads in a set of aligned DNA or protein sequences. It then displays them as graphics and highlights conserved regions in various ways.

Here is a sample session with *prettyplot*:

```
% prettyplot -resbreak=10 -boxcol -consensus -plurality=3
Displays aligned sequences, with coloring and boxing
Input sequence set: globin.msf
Graph type [x11]:

$ prettyplot globin.msf -plurality=3 -docolour
Displays aligned sequences, with coloring and boxing
Graph type [x11]:
```

Mandatory qualifiers (bold if not always prompted):

[-msf] (seqset)
: File containing a sequence alignment.

-graph *(graph)*
: Graph type.

Optional qualifiers:

-residuesperline (integer)
: Number of residues to display on each line.

-resbreak (integer)
: Residues before a space.

-[no]ccolours (boolean)
: Color residues according to their consensus value.

-cidentity (string)
: Color in which to display identical residues (default is RED).

-csimilarity (string)
: Color in which to display similar residues (default is GREEN).

-cother (string)
: Color in which to display other residues (BLACK).

-docolour (boolean)
: Color residues by table: oily, amide etc.

-[no]title (boolean)
: Do not display the title.

-shade (string)
: Set to BPLW for normal shading. Setting pair equal to 1.5,1.0,0.5 is equivalent to setting shade to BPLW.

  Residue score Color:

  1.5 or over—BLACK (B)

  1.0 to 1.5—BROWN (P)

  0.5 to 1.0—WHEAT (L)

  Under 0.5—WHITE (W)

  The only four letters allowed are B,P,L, and W (in any order).

-pair (string)
: Values representing identical, similar, or related.

-identity (integer)
: Only match those which are identical in all sequences.

-[no]box (boolean)
: Display prettyboxes.

-boxcol (boolean)
: Color the background of the boxes.

-boxcolval (string)
: Color to use for background (default is GREY).

-[no]name (boolean)
: Display the sequence names.

-maxnamelen (integer)
: Margin size for the sequence name.

-[no]number (boolean)
: Display the residue number.

-[no]listoptions (boolean)
: Display the date and options used.

-plurality (float)
: Plurality check value (total weight divided by 2).

-consensus (boolean)
: Display the consensus.

EMBOSS

-[no]collision (boolean)
: Allow collisions in calculating consensus.

-alternative (integer)
: Use alternative collisions routine:

  0 Normal collision check. (default)

  1 Checks identical scores with the max score found. If any other residue matches the identical score, a collision has occurred.

  2 If another residue has a greater than or equal to matching score and these do not match, a collision has occurred.

  3 Checks all those not in the current consensus. If any of these gives a top score for matching or identical scores, a collision has occurred.

-matrixfile (matrix)
: This is the scoring matrix file used when comparing sequences. By default, it is the file *EBLOSUM62* (for proteins) or the file *EDNAFULL* (for nucleic sequences). These files are found in the *data* directory of the EMBOSS installation.

-showscore (integer)
: Print residue scores.

-portrait (boolean)
: Set page to portrait.

Advanced qualifiers:

-data (boolean)
: Boolean value (no help text)

---

prettyseq

prettyseq writes out a well-formatted display of the sequence with the translation (within specified ranges) displayed beneath it.

Here is a sample session with *prettyseq*:

```
% prettyseq
Output sequence with translated ranges
Input sequence: embl:paamir
Range(s) to translate [1-2167]: 135-1292
Output file [paamir.prettyseq]:
```

Mandatory qualifiers:

[-sequence] (sequence)
: Sequence USA.

-range (range)
: Range(s) to translate.

[-outfile] (outfile)
: Output filename.

Optional qualifiers:

-[no]ruler (boolean)
: Add a ruler.

-[no]plabel (boolean)
Number translations.

-[no]nlabel (boolean)
Number DNA sequence.

Advanced qualifiers:

-cfile (codon)
Codon usage file.

-width (integer)
Width of screen.

primersearch

primersearch reads in primer pairs from an input file and searches them against sequence(s) specified by the user. Each of the primers in a pair is searched against the sequence, and potential amplimers are reported.

Here is a sample session with *primersearch*:

```
% primersearch embl:Z52466
Searches DNA sequences for matches with primer pairs
Primer file: primers
Output file [hsa203yc1.primersearch]: stdout
Allowed percent mismatch [0]:

Primer name D1S243

Primer name D1S468

Primer name D1S2845

Primer name D1S1608

Primer name D1S2893

Primer name D1S2660
Amplimer 1
        Sequence: HSA203YC1 Z52466
        H.sapiens (D1S2660) DNA segment containing (CA) repeat; clone
AFMa203yc1; single read.
        CACACATGCACATGCAC hits forward strand at 27 with 0 mismatches
        AGTGACACCAGCAGGG hits reverse strand at [103] with 0 mismatches
        Amplimer length: 261 bp
```

Here we run the same example, but allow a 20% mismatch between the primers and the sequence:

```
unix % primersearch embl:Z52466
Searches DNA sequences for matches with primer pairs
Primer file: primers
Output file [hsa203yc1.primersearch]: stdout
Allowed percent mismatch [0]: 20

Primer name D1S243
```

```
Primer name D1S468

Primer name D1S2845

Primer name D1S1608

Primer name D1S2893

Primer name D1S2660
Amplimer 1
        Sequence: HSA203YC1 Z52466
        H.sapiens (D1S2660) DNA segment containing (CA) repeat; clone
AFMa203yc1; single read.
        CACACATGCACATGCAC hits forward strand at 49 with 2 mismatches
        AGTGACACCAGCAGGG hits reverse strand at [103] with 0 mismatches
        Amplimer length: 239 bp
Amplimer 2
        Sequence: HSA203YC1 Z52466
        H.sapiens (D1S2660) DNA segment containing (CA) repeat; clone
AFMa203yc1; single read.
        CACACATGCACATGCAC hits forward strand at 27 with 0 mismatches
        AGTGACACCAGCAGGG hits reverse strand at [103] with 0 mismatches
        Amplimer length: 261 bp
```

Here is an example of running with a file containing a list of sequences.

```
% primersearch @seqs.list
Searches DNA sequences for matches with primer pairs
Primer file: primers
Output file [hs214yg7.primersearch]: stdout
Allowed percent mismatch [0]:

Primer name D1S243
Amplimer 1
        Sequence: HS214YG7 Z16979
        H. sapiens (D1S243) DNA segment containing (CA) repeat; clone
AFM214yg7; single read.
        CACACAGGCTCACATGCC hits forward strand at 122 with 0 mismatches
        GCTCCAGCGTCATGGACT hits reverse strand at [36] with 0 mismatches
        Amplimer length: 162 bp

Primer name D1S468
Amplimer 1
        Sequence: HS280WE5 Z23994
        H. sapiens (D1S468) DNA segment containing (CA) repeat; clone
AFM280we5; single read.
        AATTAACCGTTTTGGTCCT hits forward strand at 47 with 0 mismatches
        GCGACACACACTTCCC hits reverse strand at [96] with 0 mismatches
        Amplimer length: 185 bp

Primer name D1S2845
Amplimer 1
        Sequence: HS344WE9 Z51474
        H.sapiens (D1S2845) DNA segment containing (CA) repeat; clone
AFM344we9; single read.
```

```
        CCAAAGGGTGCTTCTC hits forward strand at 29 with 0 mismatches
        GTGGCATTCCAACCTC hits reverse strand at [157] with 0 mismatches
        Amplimer length: 201 bp

Primer name D1S1608
Amplimer 1
        Sequence: HS829186 G07829
        human STS CHLC.GATA49A06.P15262 clone GATA49A06, sequence tagged
site.
        GATGGCTTTTGGGGACTATT hits forward strand at 13 with 0 mismatches
        CACTGAGCCAAGTGACACAG hits reverse strand at [92] with 0 mismatches
        Amplimer length: 270 bp

Primer name D1S2893
Amplimer 1
        Sequence: HS123XC3 Z50993
        H.sapiens (D1S2893) DNA segment containing (CA) repeat; clone
AFM123xc3; single read.
        AAAACATCAACTCTCCCCTG hits forward strand at 5 with 0 mismatches
        CTCAAACCCCAATAAGCCTT hits reverse strand at [3] with 0 mismatches
        Amplimer length: 215 bp

Primer name D1S2660
Amplimer 1
        Sequence: HSA203YC1 Z52466
        H.sapiens (D1S2660) DNA segment containing (CA) repeat; clone
AFMa203yc1; single read.
        CACACATGCACATGCAC hits forward strand at 27 with 0 mismatches
        AGTGACACCAGCAGGG hits reverse strand at [103] with 0 mismatches
        Amplimer length: 261 bp
```

Mandatory qualifiers:

[-sequences] (seqall)
    Sequence database USA.

[-primers] (infile)
    Primer file.

[-mismatchpercent] (integer)
    Allowed percent mismatch.

[-out] (outfile)
    Output filename.

printsextract

printsextract preprocesses the *PRINTS* database for use with the program *PSCAN*.

Here is a sample session with *printsextract*:

```
% printsextract
Full pathname of PRINTS.DAT: /data/prints/prints.dat
```

Mandatory qualifiers:

[-inf] (infile)
    Full pathname of *prints.dat*.

EMBOSS

profgen

profgen generates various profiles for each alignment in a directory.

This is part of Jon Ison's protein structure analysis package. This package is still being developed. Please ignore this program until further details can be documented. All further queries should go to Jon Ison (*jison@hgmp.mrc.ac.uk*).

Here is a sample session with *profgen*:

```
% profgen
```

Mandatory qualifiers (bold if not always prompted):

-infpath (string)
Location of sequence alignment files (input).

-infextn (string)
Extension of sequence alignment files.

-type (menu)
Select type.

-threshold *(integer)*
Enter threshold reporting percentage.

-datafile *(matrixf)*
Scoring matrix.

-open *(float)*
Gap opening penalty.

-extension *(float)*
Gap extension penalty.

-smpfpath *(string)*
Location of simple profile files (output).

-smpfextn *(string)*
Extension of simple profile files.

-gbpfpath *(string)*
Location of Gribskov profile files (output).

-gbpfextn *(string)*
Extension of Gribskov profile files.

-hnpfpath *(string)*
Location of Henikoff profile files (output).

-hnpfextn *(string)*
Extension of Henikoff profile files.

profit

profit takes a simple frequency matrix produced by *prophecy* and uses it to find matches in the input sequence(s) you are searching.

Here is a sample session with *profit*:

My aligned set of sequences:

```
% more m.seq
>one
DEVGGEALGRLLVVYPWTQR
```

```
>two
DEVGREALGRLLVVYPWTQR
>three
DEVGGEALGRILVVYPWTQR
>four
DEVGGEAAGRVLVVYPWTQR
```

Make a simple frequency matrix using *prophecy*:

```
% prophecy
Creates matrices/profiles from multiple alignments
Input sequence set: m.seq
Profile type
         F : Frequency
         G : Gribskov
         H : Henikoff
Select type [F]:
Enter a name for the profile [mymatrix]:
Enter threshold reporting percentage [75]:
Output file [outfile.prophecy]:
```

Search using *profit*:

```
% profit
Scan a sequence or database with a matrix or profile
Profile or matrix file: outfile.prophecy
Input sequence(s): sw:*
Output file [outfile.profit]:
```

Mandatory qualifiers:

[-infile] (infile)
: Profile or matrix file.

[-sequence] (seqall)
: Sequence database USA.

[-outfile] (outfile)
: Output filename.

prophecy

prophecy creates a profile matrix file from a nucleic acid or a protein sequence alignment.

Here is a sample session with *prophecy*:

```
% prophecy
Creates matrices/profiles from multiple alignments
Input sequence set: globins.msf
Profile type
         F : Frequency
         G : Gribskov
         H : Henikoff
Select type [F]:
Enter a name for the profile [mymatrix]: globins
Enter threshold reporting percentage [75]:
Output file [globins.prophecy]:
```

Mandatory qualifiers (bold if not always prompted):

[-sequence] (seqset)
: Sequence set USA.

-type (menu)
: Select type.

-name (string)
: Enter a name for the profile.

-threshold *(integer)*
: Enter threshold reporting percentage.

-datafile *(matrixf)*
: Scoring matrix.

-open *(float)*
: Gap opening penalty.

-extension *(float)*
: Gap extension penalty.

[-outf] (outfile)
: Output filename.

prophet

prophet finds matches between a GRIBSKOV or HENIKOFF profile produced by *prophecy* and one or more sequences.

Here is a sample session with *prophet*:

The sequence alignment looks like:

```
% more m.seq
>one
DEVGGEA-GRLLVVYPWTQR
>two
DEVGREALGRLL-VYPWTQR
>three
DEVGGEALGRILVVY-WTQR
>four
DEVGGEAAGRVLVVYPWTQR
```

Create the profile:

```
% prophecy
Creates matrices/profiles from multiple alignments
Input sequence set: m.seq
Profile type
         F : Frequency
         G : Gribskov
         H : Henikoff
Select type [F]: g
Enter a name for the profile [mymatrix]:
Scoring matrix [Epprofile]:
Gap opening penalty [3.0]:
Gap extension penalty [0.3]:
Output file [outfile.prophecy]:
```

Now do the search.

```
% prophet
Gapped alignment for profiles
Input sequence(s): sw:*
Profile or matrix file: outfile.prophecy
Gap opening coefficient [1.0]:
Gap extension coefficient [0.1]:
Output file [100k_rat.prophet]:
```

Mandatory qualifiers:

[-sequence] (seqall)
: Sequence database USA.

[-infile] (infile)
: Profile or matrix file.

-gapopen (float)
: Gap opening coefficient.

-gapextend (float)
: Gap extension coefficient.

[-outfile] (outfile)
: Output filename.

prosextract

prosextract builds the *PROSITE* motif database for *patmatmotifs* to search.

Here is a sample session with *prosextract*:

```
% prosextract
Extracting ID, AC & PA lines from the Prosite motif Database.
Enter name of prosite directory: data/PROSITE

% more prosite.lines
ASN_GLYCOSYLATION PS00001
N-glycosylation
N-{P}-[ST]-{P}
^N[^P][ST][^P]

CAMP_PHOSPHO_SITE PS00004
cAMP-
[RK](2)-x-[ST]
^[RK]{2}[^BJOUXZ][ST]

PKC_PHOSPHO_SITE PS00005
Protein
[ST]-x-[RK]
^[ST][^BJOUXZ][RK]

CK2_PHOSPHO_SITE PS00006
Casein
[ST]-x(2)-[DE]
^[ST][^BJOUXZ]{2}[DE]

etc.......
```

The output files named after the *prosite* access numbers can also be seen in the *prosite* directory. This files are automatically created after *prosextract* is run.

Mandatory qualifiers:

[-infdat] (string)
Enter name of *PROSITE* directory.

pscan

pscan scans proteins using *PRINTS*. The home web page of the *PRINTS* database is *http://www.bioinf.man.ac.uk/dbbrowser/PRINTS/*.

Here is a sample session with *pscan*:

```
% pscan
Scans proteins using PRINTS
Input sequence(s): sw:OPSD_HUMAN
Minimum number of elements per fingerprint [2]:
Maximum number of elements per fingerprint [20]:
Output file [opsd_human.pscan]:
```

Mandatory qualifiers:

[-sequence] (seqall)
Sequence database USA.

-emin (integer)
Minimum number of elements per fingerprint.

-emax (integer)
Maximum number of elements per fingerprint.

[-outfile] (outfile)
Output filename.

psiblast

psiblast runs PSI-BLAST given *scopalign* alignments.

This is part of Jon Ison's protein structure analysis package. This package is still being developed. Please ignore this program until further details can be documented. All further queries should go to Jon Ison (*jison@hgmp.mrc.ac.uk*).

Here is a sample session with *psiblast*:

```
% psiblasts
```

Mandatory qualifiers:

[-align] (string)
Location of alignment files for input.

[-alignextn] (string)
File extension of alignment files.

[-niter] (integer)
Number of PSI-BLAST iterations.

[-evalue] (float)
Threshold E-value for inclusion in family.

[-maxhits] (integer)
: Maximum number of hits.

[-submatrix] (string)
: Residue substitution matrix.

[-families] (outfile)
: Name of families file for output.

[-logf] (outfile)
: Name of log file for the build.

---

rebaseextract

rebaseextract derives recognition site and cleavage information from the *withrefm* file of an REBASE distribution. It creates three files in the *EMBOSS* data subdirectory *REBASE*: a pattern file, reference file, and supplier file.

Here is a sample session with *rebaseextract*:

```
% rebaseextract
Full pathname of WITHREFM: /data/rebase/withrefm.904
```

Mandatory qualifiers:

[-inf] (infile)
: Full pathname of *WITHREFM*.

---

recoder

recoder scans a given nucleotide sequence for restriction sites. It reports single base positions in the restriction pattern which when mutated remove the restriction site while maintaining the same translation (in frame 1 of the input sequence).

Here is a sample session with *recoder*:

```
% recoder
Find and remove restriction sites but maintain the same translation
Input sequence: em:hsfau
Comma separated enzyme list [all]: EcoRII
Output file [hsfau.recoder]:
```

Mandatory qualifiers:

[-seq] (sequence)
: Nucleic acid sequence.

-enzymes (string)
: Comma-separated enzyme list

[-outf] (outfile)
: Results filename

Advanced qualifiers:

-sshow (boolean)
: Display untranslated sequence.

-tshow (boolean)
: Display translated sequence.

redata

redata searches the REBASE database for information on a specified restriction enzyme.

Here is a sample session with *redata*:

```
% redata
Search REBASE for enzyme name, references, suppliers etc.
Restriction enzyme name [BamHI]:
Output file [outfile.redata]:
```

Mandatory qualifiers:

[-enzyme] (string)
: Enter the name of the restriction enzyme you want to examine. The names often contain an "I". This is a capital letter "i", not a numeric "1" or the letter "l". The names are case-independent (`AaeI` is the same as `aaei`).

[-outfile] (outfile)
: Output filename.

Advanced qualifiers:

-[no]isoschizomers (boolean)
: Show other enzymes with this specificity (isoschizomers).

-[no]references (boolean)
: Show references.

-[no]suppliers (boolean)
: Show suppliers.

remap

remap uses REBASE data to find the recognition sites and/or cut sites of restriction enzymes in a nucleic acid sequence. It also displays the cut sites on both strands by default. It will optionally also display the translation of the sequence.

Here is a sample session with *remap*. We only look at a small section of the sequence to save space:

```
% remap -notran -sbeg 1 -send 60
Display a sequence with restriction cut sites, translation etc..
Input sequence(s): embl:eclac
Output file [eclac.remap]:
Comma separated enzyme list [all]: taqi,bsu6i,acii,bsski
Minimum recognition site length [4]:
```

Here is an example where all enzymes in the REBASE database are used:

```
% remap -notran -sbeg 1 -send 60
Display a sequence with restriction cut sites, translation etc..
Input sequence(s): embl:eclac
Output file [eclac.remap]:
Comma separated enzyme list [all]:
Minimum recognition site length [4]:
```

Mandatory qualifiers:

[-sequence] (seqall)
Sequence database USA.

-enzymes (string)
The argument `all` reads all enzyme names from the REBASE database. You can specify enzymes by giving their names with commas between them, such as: `HincII,hinfI,ppiI,hindiii`. This command is not case-sensitive. You may also use the data from file containing enzyme names by prepending the name of the file you want to use with an `@` character; for example, `@enz.list`. Blank lines and lines starting with a comment tag (`#` or `!`) within the file are ignored; all other lines are concatenated together with a comma and treated as the list of enzymes to search for. A file containing enzyme names might look like this:

```
! my enzymes
HincII, ppiII
! other enzymes
hindiii
HinfI
PpiI
```

-sitelen (integer)
Minimum recognition site length.

[-outfile] (outfile)
If you enter the name of a file here, this program will write the sequence details into that file.

Optional qualifiers:

-mincuts (integer)
Minimum cuts per restriction enzyme.

-maxcuts (integer
Maximum cuts per restriction enzyme.

-single (boolean)
Force single-site-only cuts.

-[no]blunt (boolean)
Allow blunt end cutters.

-[no]sticky (boolean)
Allow sticky end cutters.

-[no]ambiguity (boolean)
Allow ambiguous matches.

-plasmid (boolean)
Allow circular DNA.

-[no]commercial (boolean)
Only enzymes with suppliers.

-table (menu)
Code to use. See the *fuzztran* description for codes.

-[no]cutlist (boolean)
List the enzymes to cut.

-flatreformat (boolean)
Display restriction enzyme sites in flat format.

-[no]limit (boolean)
Limits reports to one isoschizomer.

-preferred (boolean)
Report preferred isoschizomers.

Advanced qualifiers:

-[no]translation (boolean)
Display translation.

-[no]reverse (boolean)
Display cut sites and translation of reverse sense.

-orfminsize (integer)
Minimum size of Open Reading Frames (ORFs) to display in the translations.

-uppercase (range)
Regions to put in uppercase. If no regions are specified, the sequence case is left alone. A set of regions is specified by a set of pairs of positions. The positions are integers. They are separated by any non-digit, non-alpha character. Examples of region specification: `24-45, 56-78`, `1:45, 67=99;765..888`, `1,5,8,10,23,45,57,99`.

-highlight (range)
Regions to color if formatting in HTML. If no regions are specified, the sequence is left alone. A set of regions is specified by a set of pairs of positions. The positions are integers. They are followed by any valid HTML font color. Examples of region specifications:

```
24-45 blue 56-78 orange
1-100 green 120-156 red
```

A file of ranges to color (one range per line) can be specifed as `@filename`.

-threeletter (boolean)
Display protein sequences in three-letter code.

-number (boolean)
Number the sequences.

-width (integer)
Width of sequence to display.

-length (integer)
Line length of page (0 for indefinite length).

-margin (integer)
Margin around sequence for numbering.

-[no]name (boolean)
Set this to false if you do not want to display the ID name of the sequence.

-[no]description (boolean)
Set this to false if you do not want to display the description of the sequence.

-offset (integer)
Offset to start numbering the sequence from.

-html (boolean)
Use HTML formatting.

restover

restover takes a specified sequence and a short sequence of a cut-site overhang and searches the REBASE database for matching enzymes that create the desired overhang sequence when they cut the input sequence.

Here is a sample session with *restover*:

```
% restover
Finds restriction enzymes that produce a specific overhang
Input sequence(s): em:hsfau
Overlap sequence: cg
Output file [hsfau.restover]:
```

EMBOSS

Mandatory qualifiers:

[-sequence] (seqall)
: Sequence database USA.

[-seqcomp] (string)
: Overlap sequence.

[-outfile] (outfile)
: Output filename.

Advanced qualifiers:

-min (integer)
: Minimum cuts per restriction enzyme.

-max (integer)
: Maximum cuts per restriction enzyme.

-single (boolean)
: Force single site only cuts

-threeprime (boolean)
: 3' overhang? (else 5') e.g., BamHI has CTAG as a 5' overhang, and ApaI has CCGG as 3' overhang.

-[no]blunt (boolean)
: Allow blunt end cutters.

-[no]sticky (boolean)
: Allow sticky end cutters.

-[no]ambiguity (boolean)
: Allow ambiguous matches.

-plasmid (boolean)
: Allow circular DNA.

-[no]commercial (boolean)
: Only enzymes with suppliers.

-datafile (string)
: Alternative restriction enzyme data file.

-html (boolean)
: Create HTML output.

-[no]limit (boolean)
: Limits reports to one isoschizomer.

-preferred (boolean)
: Report preferred isoschizomers.

-alphabetic (boolean)
: Sort output alphabetically.

-fragments (boolean)
: Show fragment lengths.

-name (boolean)
: Show sequence name.

restrict

restrict uses the REBASE database of restriction enzymes to predict cut sites in a DNA sequence.

Here is a sample session with *restrict*:

```
% restrict
Finds restriction enzyme cleavage sites
Input sequence(s): embl:hsfau
Minimum recognition site length [4]:
Comma separated enzyme list [all]:
Output file [hsfau.restrict]:
```

Mandatory qualifiers:

[-sequence] (seqall)
: Sequence database USA.

-sitelen (integer)
: Minimum recognition site length.

-enzymes (string)
: The argument `all` reads all enzyme names from the REBASE database. You can specify enzymes by giving their names with commas between them, such as: `HincII,hinfI,ppiI,hindiii`. This command is not case-sensitive. You may also use the data from file containing enzyme names by prepending the name of the file you want to use with an `@` character; for example, `@enz.list`. Blank lines and lines starting with a comment tag (`#` or `!`) within the file are ignored; all other lines are concatenated together with a comma and treated as the list of enzymes to search for. A file containing enzyme names might look like this:

```
! my enzymes
HincII, ppiII
! other enzymes
hindiii
HinfI
PpiI
```

[-outfile] (report)
: Output report filename.

Advanced qualifiers:

-min (integer)
: Minimum cuts per restriction enzyme.

-max (integer)
: Maximum cuts per restriction enzyme.

-single (boolean
: Force single site only cuts.

-[no]blunt (boolean)
: Allow blunt end cutters.

-[no]sticky (boolean)
: Allow sticky end cutters.

-[no]ambiguity (boolean)
: Allow ambiguous matches.

-plasmid (boolean)
: Allow circular DNA.

-[no]commercial (boolean)
: Only enzymes with suppliers.

-datafile (string)
: Alternative restriction enzyme data file.

-[no]limit (boolean)
: Limits reports to one isoschizomer.

-preferred (boolean)
: Report preferred isoschizomers.

-alphabetic (boolean)
: Sort output alphabetically.

-fragments (boolean)
: Show fragment lengths.

-name (boolean)
: Show sequence name.

---

revseq

revseq takes a sequence and outputs the reverse complement (also known as the anti-sense or reverse sense) sequence.

To create the reverse sense of the sequence *rcr.seq* and output to the file *rcr.rev*:

```
% revseq rcr.seq rcr.rev
```

To create the complement of the sequence *rcr.seq* and output to the file *rcr.rev*:

```
% revseq rcr.seq rcr.rev -norev
```

To create the reverse of the sequence *rcr.seq* and output to the file *rcr.rev*:

```
% revseq rcr.seq rcr.rev -nocomp
```

Mandatory qualifiers:

[-sequence] (seqall)
: Sequence database USA.

[-outseq] (seqoutall)
: Output sequence(s) USA.

Advanced qualifiers:

-[no]reverse (boolean)
: Set this to be false if you do not want to reverse the output sequence.

-[no]complement (boolean)
: Set this to be false if you do not want to complement the output sequence.

scopalign

scopalign generate alignments for families in a scop classification file by using STAMP.

Here is a sample session with *scopalign*:

```
% scopalign
```

Mandatory qualifiers:

[-scopf] (infile)
Name of scop classification file (EMBL format input).

[-path] (string)
Location of scop structure-based sequence alignment files (output).

[-extn] (string)
Extension of scop structure-based sequence alignment files (output).

[-pathc] (string)
Location of scop structure alignment files (output).

[-extnc] (string)
Extension of scop structure alignment files (output).

scope

scope writes the SCOP classification to an EMBL-like format file.

Here is a sample session with *scope*:

```
% scope
Convert raw scop classification file to embl-like format
Name of scop file for input (raw format) [scop.orig]: /data/scop/scop.orig
Name of scop file for output (embl-like format) [Escop.dat]: Escop.test
```

Mandatory qualifiers:

[-infile] (infile)
Name of scop file for input (raw format).

[-outfile] (outfile)
Name of scop file for output (EMBL-like format).

scopnr

scopnr removes redundant domains from a scop classification file.

This is part of Jon Ison's protein structure analysis package. This package is still being developed. Please ignore this program until further details can be documented. All further queries should go to Jon Ison (*jison@hgmp.mrc.ac.uk*).

Here is a sample session with *scopnr*:

```
% scopnr
```

Mandatory qualifiers (bold if not always prompted):

[-scopin] (infile)
Name of scop classification file (EMBL format input).

-mode (menu)
Select number.

-thresh *(float)*
The percentage sequence identity redundancy threshold.

-threshlow *(float)*
The percentage sequence identity redundancy threshold (lower limit).

-threshup *(float)*
The percentage sequence identity redundancy threshold (upper limit).

[-scopout] (outfile)
Name of non-redundant scop classification file (EMBL format output).

-errf (outfile)
Name of log file for the build.

Optional qualifiers:

-datafile (matrixf)
This is the scoring matrix file used when comparing sequences.

-gapopen (float)
The gap insertion penalty is the score taken away when a gap is created. The best value depends on the choice of comparison matrix. The default value assumes you are using the EBLOSUM62 matrix for protein sequences, and the EDNAFULL matrix for nucleotide sequences.

-gapextend (float)
The gap extension penalty is added to the standard gap penalty for each base or residue in the gap. This is how long gaps are penalized. You can usually expect a few long gaps rather than many short gaps, so the gap extension penalty should be lower than the gap penalty. An exception occurs when one or both sequences are single reads with possible sequencing errors, in which case you should expect many single base gaps. You can obtain this result by setting the gap open penalty to zero (or a very low value) and using the gap extension penalty to control gap scoring.

---

scopparse

scopparse converts raw scop classification files to a file in EMBL-like format.

This is part of Jon Ison's protein structure analysis package. This package is still being developed. Please ignore this program until further details can be documented. All further queries should go to Jon Ison (*jison@hgmp.mrc.ac.uk*).

Here is a sample session with *scopparse*:

```
% scopparse
```

Mandatory qualifiers:

[-infilea] (infile)
Name of scop classification file (raw format *dir.cla.scop.txt_X.XX* for input).

[-infileb] (infile)
Name of scop description file (raw format *dir.des.scop.txt_X.XX* for input).

[-outfile] (outfile)
Name of scop classification file (EMBL format output).

scoprep

scoprep reorders a scop classificaition file so that the representative structure of each family is given first.

This is part of Jon Ison's protein structure analysis package. This package is still being developed. Please ignore this program until further details can be documented. All further queries should go to Jon Ison (*jison@hgmp.mrc.ac.uk*).

Here is a sample session with *scoprep*:

```
% scoprep
```

Mandatory qualifiers:

[-scopin] (infile)
: Name of scop classification file (EMBL format input).

[-scopout] (outfile)
: Name of scop classification file (EMBL format output).

scopreso

scopreso removes low resolution domains from a scop classification file.

This is part of Jon Ison's protein structure analysis package. This package is still being developed. Please ignore this program until further details can be documented. All further queries should go to Jon Ison (*jison@hgmp.mrc.ac.uk*).

Here is a sample session with *scopreso*:

```
% scopreso
```

Mandatory qualifiers:

[-cpdbpath] (string)
: Location of domain coordinate files for input (EMBL-like format).

[-cpdbextn] (string)
: Extension of coordinate files (EMBL-like format).

[-scopinf] (infile)
: Name of SCOP data file for input (EMBL-like format).

-threshold (float)
: Threshold for inclusion (Angstroms).

[-scopoutf] (outfile)
: Name of SCOP data file for output.

scopseqs

scopseqs adds PDB and SWISS-PROT sequence records to a scop classification file.

This is part of Jon Ison's protein structure analysis package. This package is still being developed. Please ignore this program until further details can be documented. All further queries should go to Jon Ison (*jison@hgmp.mrc.ac.uk*).

Here is a sample session with *scopseqs*:

```
% scopseqs
```

Mandatory qualifiers:

[-scopin] (infile)
: Name of scop file for input (EMBL-like format).

[-pdbtosp] (infile)
: Name of the pdbcodes to SWISS-PROT indexing (EMBL-like format).

[-dpdb] (string)
: Location of clean domain coordinate files for input (EMBL-like format).

[-extn] (string)
: File extension of clean domain coordinate files.

[-scopout] (outfile)
: Name of processed file for output (EMBL-like format).

[-errf] (outfile)
: Name of log file for the build.

Advanced qualifiers:

-datafile (matrixf)
: This is the scoring matrix file used when comparing sequences.

-gapopen (float)
: The gap insertion penalty is the score taken away when a gap is created. The best value depends on the choice of comparison matrix. The default value assumes you are using the EBLOSUM62 matrix for protein sequences, and the EDNAFULL matrix for nucleotide sequences.

-gapextend (float)
: The gap extension penalty is added to the standard gap penalty for each base or residue in the gap. This is how long gaps are penalized. You can usually expect a few long gaps rather than many short gaps, so the gap extension penalty should be lower than the gap penalty. An exception occurs when one or both sequences are single reads with possible sequencing errors, in which case you should expect many single base gaps. You can obtain this result by setting the gap open penalty to zero (or a very low value) and using the gap extension penalty to control gap scoring.

seealso

seealso takes the name of an existing program in EMBOSS (or a program in one the associated EMBASSY packages) and gives a list of the programs which share some functionality with the existing program.

Here is a sample session with *seealso*:

```
% seealso
Program to search for: matcher
SEE ALSO
seqmatchall       Does an all-against-all comparison of a set of sequences
supermatcher      Finds a match of a large sequence against one or more
                  sequences
water             Smith-Waterman local alignment
wordmatch         Finds all exact matches of a given size between 2 sequences
```

Mandatory qualifiers:

[-search] (string)

Enter the name of an EMBOSS program.

Optional qualifiers (bold if not always prompted):

-explode (boolean)

The groups EMBOSS applications belong to have two forms: exploded and non-exploded. The exploded group names are more numerous and often more vaguely phrased than the non-exploded ones. The exploded names are formed from definitions of the group names that start in the form `NAME1:NAME2` and are then expanded into many combinations of the names, such as: `NAME1, NAME2, NAME1 NAME2, NAME2 NAME1`. The non-expanded names are simply listed as: `NAME1 NAME2`. Using exploded group names finds many more programs that share at least some of the exploded names. More programs will be reported as sharing a similar function if you use use non-exploded names.

-outfile (outfile)

If you enter the name of a file here, this program writes the program names and brief descriptions into that file.

-html (boolean)

If you are sending the output to a file, this qualifier formats it for display as a table in a WWW document.

-prelink *(string)*

If you are outputting to a file in HTML format, you can make the program names into hyperlinks by setting this qualifier to be the first half of the URL. If you want the program name to be a hyperlink to the URL *http://www.uk.embnet.org/Software/EMBOSS/Apps/programname.html*, enter `http://www.uk.embnet.org/Software/EMBOSS/Apps/` to set the first half of the URL before the program name.

-postlink *(string)*

If you are outputting to a file in HTML format, you can make the program names into hyperlinks by setting this qualifier to be the second half of the URL. If you want the program name to be a hyperlink to the UR: *http://www.uk.embnet.org/Software/EMBOSS/Apps/programname.html*, enter `.html` to set the second half of the URL after the program name.

-groups (boolean)

If you use this option, only the group names will output to the file.

Advanced qualifiers:

-[no]emboss (boolean)

If you use this option, EMBOSS program documentation is searched. If this option is set to be false, only the EMBASSY programs are searched (if the -`embassy` option is true). EMBASSY programs are not strictly part of EMBOSS, but use the same code libraries and share the same look and feel. These programs are generally developed by people who want their programs to be outside of the GNU Public License scheme, i.e., they may want to charge a license fee.

-[no]embassy (boolean)

If you use this option, EMBASSY program documentation is searched. If this option is set to false, only the EMBOSS programs will be searched (if the `-emboss` option is true). EMBASSY programs are not strictly part of EMBOSS, but use the same code libraries and share the same look and feel, but are generally developed by people who want the programs to be outside of the GNU Public License scheme, i.e.,. they may want to charge a license fee.

-colon (boolean)
: The groups EMBOSS applications belong to may have up to two levels For example, the primary group ALIGNMENT has several sub-groups, or second-level groups: CONSENSUS, DIFFERENCES, DOT PLOTS, GLOBAL, LOCAL, and MULTIPLE. To aid programs that parse the output of *seealso* (and require the names of these subgroups), a colon is placed between the first and second level of the group name if this option is true.

seqalign

seqalign generate extended alignments for families in a scop families file by using ClustalW with seed alignments.

This is part of Jon Ison's protein structure analysis package. This package is still being developed. Please ignore this program until further details can be documented. All further queries should go to Jon Ison (*jison@hgmp.mrc.ac.uk*).

Here is a sample session with *seqalign*:

```
% seqalign
```

Mandatory qualifiers:

[-inpath] (string)
: Location of scop alignment files (input).

[-extn] (string)
: Extension of scop alignment files.

-scopin (infile)
: Name of scop hits file (input).

[-outpath] (string)
: Location of extended alignment files (Escop format output).

[-outextn] (string)
: Extension of extended alignment files.

seqmatchall

seqmatchall does an all-against-all pairwise comparison of words (fragments of the sequences of a specified fixed size) in a set of sequences and finds regions of similarity within any two sequences.

Here is a sample session with *seqmatchall*. We use an increased word size to avoid accidental matches:

```
% seqmatchall
Does an all-against-all comparison of a set of sequences
Input sequence set: embl:eclac*
Word size [4]: 15
Output file [outfile.seqmatchall]:
```

Mandatory qualifiers:

[-sequence] (seqset)
: Sequence set USA.

-wordsize (integer)
: Word size.

[-outfile] (outfile)
: Output filename.

---

seqnr

seqnr converts redundant database results to a non-redundant set of hits.

This is part of Jon Ison's protein structure analysis package. This package is still being developed. Please ignore this program until further details can be documented. All further queries should go to Jon Ison (*jison@hgmp.mrc.ac.uk*).

Here is a sample session with *seqnr*:

```
% seqnr
```

Mandatory qualifiers:

[-inf] (infile)
: Name of the scop hits file (input).

[-vinf] (infile)
: Name of validation file (input).

-scopin (infile)
: Name of scop classification file (EMBL format input).

-alignpath (string)
: Location of scop alignment files (input).

-alignextn (string)
: Extension of scop alignment files.

-datafile (matrixf)
: This is the scoring matrix file used when comparing sequences.

-thresh (float)
: The percentage sequence identity redundancy threshold.

-mode (menu)
: Select mode.

[-outf] (outfile)
: Name non-redundant scop hits file (output).

[-voutfname] (string)
: Name of processed validation file (output).

-logf (outfile)
: Name *seqnr* log file (output).

Advanced qualifiers:

-gapopen (float)
: The gap insertion penalty is the score taken away when a gap is created. The best value depends on the choice of comparison matrix. The default value assumes you are using the EBLOSUM62 matrix for protein sequences, and the EDNAFULL matrix for nucleotide sequences.

-gapextend (float)
: The gap extension penalty is added to the standard gap penalty for each base or residue in the gap. This is how long gaps are penalized. You can usually expect a few long gaps rather than many short gaps, so the gap extension penalty should be lower than the gap penalty. An exception occurs when one or both sequences are single reads with possible sequencing errors, in which case you should expect many single base gaps. You can obtain this result by setting the gap open penalty to zero (or a very low value) and using the gap extension penalty to control gap scoring.

seqret

seqret reads and writes (returns) sequences.

Here is a sample session with *seqret*. It is used to extract an entry from a database and write it to a file:

```
% seqret
Input sequence: embl:paamir
Output sequence [paamir.fasta]:
```

Here *seqret* is used to display the contents of the sequence on the screen:

```
% seqret
Reads and writes (returns) sequences
Input sequence: embl:paamir
Output sequence [paamir.fasta]: stdout
```

Here it is used in three different ways to write the result to a file in GCG format. Once by using the qualifier `-osformat`, once by using the format in the output USA on the command line, and once by specifying the format in the USA at the prompt:

```
% seqret -osf gcg
Reads and writes (returns) sequences
Input sequence: embl:paamir
Output sequence [paamir.gcg]:

% seqret -outseq gcg::paamir.gcg
Reads and writes (returns) sequences
Input sequence: embl:paamir


% seqret
Reads and writes (returns) sequences
Input sequence: embl:paamir
Output sequence [paamir.fasta]: gcg::paamir.gcg
```

Here *seqret* is used to produce the reverse-complement of a sequence:

```
% seqret -srev
Reads and writes (returns) sequences
Input sequence: embl:paamir
Output sequence [paamir.fasta]:
```

Here *seqret* is used to extract the bases between the positions starting at 5 and ending at 25:

```
% seqret -sbegin 5 -send 25
Reads and writes (returns) sequences
Input sequence: embl:paamir
Output sequence [paamir.fasta]:
```

Here *seqret* is used to extract the bases between the positions starting at 5 and ending at 5 bases before the end of the sequence:

```
% seqret -sbegin 5 -send -5
Reads and writes (returns) sequences
Input sequence: embl:paamir
Output sequence [paamir.fasta]:
```

Here *seqret* is used to read all entries in the database *tembl* that start with "hs" and write them to a file:

```
% seqret
Reads and writes (returns) sequences
Input sequence(s): embl:hs*
Output sequence [hs989235.fasta]:
```

Here *seqret* is used to read all entries in the database *tembl* that start with "hs" and writes them to a file. In this example, the specification is all done in the command line. To avoid confusing Unix with the * character, put backslash (\) before it:

```
% seqret embl:hs\* hs989235.fasta
Reads and writes (returns) sequences
```

Here *seqret* is used to read only the entry "hsfau" from the file *hs989235.fasta*, which contains many entries:

```
% seqret
Reads and writes (returns) sequences
Input sequence(s): hs989235.fasta:hsfau
Output sequence [hsfau.fasta]:
```

Here *seqret* is used to read all entries in the file *hs989235.fasta*, but only writes the first one of these entries out to a file:

```
% seqret -firstonly
Reads and writes (returns) sequences
Input sequence(s): hs989235.fasta
Output sequence [hs989235.fasta]: first.fasta
```

Here *seqret* is used to display the short sequence *actgatcgtg* in uppercase EMBL format on the screen:

```
% seqret -supper -osf embl asis::actgatcgtg stdout
Reads and writes (returns) sequences
ID             standard; DNA; UNC; 10 BP.
SQ   Sequence 10 BP; 2 A; 2 C; 3 G; 3 T; 0 other;
     ACTGATCGTG                                    10
//
```

To force *seqret* to both read in and write out features, use the command-line qualifier `-feature`. *seqret* does not read in features by default, resulting in slightly faster performance. If you want to read features with your sequence and write them out on output, using `-feature` will change the default behavior so that any features present in the sequence are used:

```
% seqret -feature
Reads and writes (returns) sequences
Input sequence(s): em:hs989235
Output sequence [hs989235.fasta]: embl::hs989235.embl
```

Mandatory qualifiers:

[-sequence] (seqall)
Sequence database USA.

[-outseq] (seqoutall)
Output sequence(s) USA.

Advanced qualifiers:

-feature (boolean)
Use feature information.

-firstonly (boolean)
Read one sequence and stop.

seqretsplit

seqretsplit reads and writes (returns) sequences in individual files. *seqretsplit* is exactly the same as *seqret*, except it writes each sequence to an individual file that when writing out more than one sequence. Therefore, its main use is to split a file containing multiple sequences into many files, each containing one sequence.

Here is a sample session with *seqretsplit*:

```
% seqretsplit
Reads and writes (returns) sequences in individual files
Input sequence(s): embl:hsfa11*
Output sequence [hsfa110.fasta]:
```

The specification of the output file is not used in this case. At some point this ought to change, and you will not be prompted for the output file.

Mandatory qualifiers:

[-sequence] (seqall)
Sequence database USA.

[-outseq] (seqoutall)
Output sequence(s) USA.

Advanced qualifiers:

-firstonly (boolean)
Read one sequence and stop.

seqsearch

seqsearch generate files of hits for families in a scop classification file by using PSI-BLAST with seed alignments.

This is part of Jon Ison's protein structure analysis package. This package is still being developed. Please ignore this program until further details can be documented. All further queries should go to Jon Ison (*jison@hgmp.mrc.ac.uk*).

Here is a sample session with *seqsearch*:

```
% seqsearch
```

Mandatory qualifiers:

[-escop] (infile)
Name of scop classification file (EMBL format input).

[-align] (string)
Location of scop alignment files (input).

[-alignextn] (string)
Extension of scop alignment files.

-niter (integer)
Number of PSI-BLAST iterations.

-evalue (float)
Threshold E-value for inclusion in family.

-maxhits (integer)
Maximum number of hits.

-submatrix (string)
Residue substitution matrix.

[-hits] (string)
Location of scop hits files (output).

[-hitsextn] (string)
Extension of scop hits files.

-logf (outfile)
Name of log file for the build.

---

seqsort

seqsort reads multiple files of hits and writes a non-ambiguous file of hits (scop families file) plus a validation file.

This is part of Jon Ison's protein structure analysis package. This package is still being developed. Please ignore this program until further details can be documented. All further queries should go to Jon Ison (*jison@hgmp.mrc.ac.uk*).

Here is a sample session with *seqsort*:

```
% seqsort
```

Mandatory qualifiers (bold if not always prompted):

-mode (menu)
Select mode.

-psipath *(string)*
Location of scop hits files (input).

-psiextn *(string)*
Extension of scop hits files.

-swisspath *(string)*
Location of seqwords hits files.

-swissextn *(string)*
Extension of seqwords input files.

-psifile *(string)*
: Name of file containing processed scop hits file (input).

-swissfile *(string)*
: Name of file containing processed seqwords hits file (input).

-overlap (integer)
: number of overlapping residues required for merging of two hits.

[-hitsf] (outfile)
: Name of scop hits file (output).

[-validf] (outfile)
: Name of validation file (output).

seqwords

seqwords generates a file of hits for scop families by searching SWISS-PROT with keywords.

This is part of Jon Ison's protein structure analysis package. This package is still being developed. Please ignore this program until further details can be documented. All further queries should go to Jon Ison (*jison@hgmp.mrc.ac.uk*).

Here is a sample session with *seqwords*:

```
% seqwords
```

Mandatory qualifiers:

[-keyfile] (infile)
: Name of keywords file (input).

-spfile (infile)
: Name of SWISS-PROT database (input).

[-outfile] (outfile)
: Name of seqwords hits file (output).

showalign

showalign displays an aligned set of protein or a nucleic acid sequences in a style suitable for publication.

Here is a sample session with *showalign*:

```
% showalign
Displays a multiple sequence alignment
Input sequence set: ~/align.pep
Output file [align.showalign]:
```

Mandatory qualifiers:

[-sequence] (seqset)
: The sequence alignment to be displayed.

[-outfile] (outfile)
: If you enter the name of a file, this program writes the sequence details into that file.

Optional qualifiers:

-refseq (string)

If you give the number in the alignment or the name of a sequence, it is taken as the reference sequence. The reference sequence is always shown in full and is the one against which all the other sequences are compared. If this is qualifier is set to `0`, the consensus sequence is used as the reference sequence. By default, the consensus sequence is used as the reference sequence.

-[no]bottom (boolean)

If this option is true, the reference sequence is displayed at the top and bottom of the alignment.

-show (menu)

What to show.

-order (menu)

Output order of the sequences.

-[no]similarcase (boolean)

If this option is true (when `-show` is set to `Similarities` or `Non-identities` and a residue is similar but not identical to the reference sequence residue), the residue case is changed to lowercase. If `-show` is set to `All`, non-identical, non-similar residues are changed to lowercase. If `False`, no changes are made to the case of the residues on the basis of their similarity to the reference sequence.

-matrix (matrix)

This is the scoring matrix file used when comparing sequences. By default, it is the file *EBLOSUM62* (for proteins) or the file *EDNAFULL* (for nucleic sequences). These files are found in the *data* directory of the EMBOSS installation.

-[no]consensus (boolean)

If this option is true, the consensus line is displayed at the bottom.

Advanced qualifiers:

-uppercase (range)

Regions to put in uppercase. If this is left blank, the sequence case is left alone. A set of regions is specified by a set of pairs of positions. The positions are integers. They are separated by any non-digit, non-alpha character. Examples of region specifications are:

```
24-45, 56-78
1:45, 67=99;765..888
1,5,8,10,23,45,57,99
```

-[no]number (boolean)

If this option is true, a line giving the positions in the alignment is displayed every 10 characters above the alignment.

-[no]ruler (boolean)

If this option is true, a ruler line marking every 5th and 10th character in the alignment is displayed.

-width (integer)

Width of sequence to display.

-margin (integer)
: This sets the length of the left-hand margin for sequence names. If the margin is set at `0`, no margin and no names are displayed. If the margin is set to a value less than the length of a sequence name, the sequence name is displayed truncated to the length of the margin. If the margin is set to `-1`, the minimum margin width that allows all the sequence names to be displayed in full (plus a space at the end of the name) is automatically selected.

-html (boolean)
: Use HTML formatting.

-highlight (range)
: Regions to color if formatting for HTML. If this is left blank, the sequence is left alone. A set of regions is specified by a set of pairs of positions. The positions are integers. They are followed by any valid HTML font color. Examples of region specifications are:

```
24-45 blue 56-78 orange
1-100 green 120-156 red
```

A file of ranges to color (one range per line) can be specified as `@filename`.

-plurality (float)
: Set a cut-off for the percentage of positive scoring matches below which there is no consensus. The default plurality is taken as 50% of the total weight of all sequences in the alignment.

-setcase (float)
: Sets the threshold for the scores of the positive matches above which the consensus is is uppercase, and below which the consensus is in lowercase.

-identity (float)
: Provides the ability to set the required number of identities at a position for it to give a consensus. If this is set to 100%, only columns of identities contribute to the consensus.

showdb

showdb writes out a simple table displaying the names, contents, and available ways to access the sequence databases.

Display information on the currently available databases:

```
% showdb
```

Write out the display to a file:

```
% showdb -outfile showdb.out
```

Display information on one explicit database:

```
% showdb -database swissprot
```

Display information on the databases formatted for inclusion in HTML:

```
% showdb -html
```

Display protein databases only:

```
% showdb -nonucleic
```

Display the information with no headings:

```
% showdb -noheading
```

Display just a list of the available database names:

```
% showdb -noheading -notype -noid -noquery -noall -nocomment -auto
```

Display only the names and types:

```
% showdb -only -type
```

Optional qualifiers:

-database (string)
: Name of a single database to give information on.

-html (boolean)
: Format output as an HTML table.

-[no]protein (boolean)
: Display protein databases.

-[no]nucleic (boolean)
: Display nucleic acid databases.

-release (boolean)
: Display "release" column.

-outfile (outfile)
: If you enter the name of a file here, this program writes the database details into that file.

Advanced qualifiers:

-only (boolean)
: This is a way of shortening the command line if you only want a few things to be displayed. Instead of specifying: `-nohead -notype -noid -noquery -noall` to get only the comment output, you can specify `-only -comment`.

-heading (boolean)
: Display column headings.

-type (boolean)
: Display "type" column.

-id (boolean)
: Display "id" column.

-query (boolean)
: Display "qry" column.

-all (boolean)
: Display "all" column.

-comment (boolean)
: Display "comment" column.

showfeat

showfeat reads a protein or nucleic sequence and its feature table and writes a text representation of the features to standard output.

Here is a sample session with *showfeat*. The feature table is specified as a `-ufo` qualifier (uniform feature object)—in this case, a file containing an EMBL feature table:

```
% showfeat em:paamir
Show features of a sequence.
Output file [paamir.showfeat]: stdout
PAAMIR
Pseudomonas aeruginosa amiC and amiR gene for aliphatic amidase regulation
|==============================================| 2167
>                                                misc_feature
|----------------------------------------------> source
>                                                promoter
|>                                               promoter
  >                                              RBS
  |---------------------------->                 CDS
             |>                                  misc_feature
                    |----->                      variation
                             >                   conflict
                             |-------------->    CDS
```

Mandatory qualifiers:

[-sequence] (seqall)
: Sequence database USA.

[-outfile] (outfile)
: If you enter the name of a file here, this program will write the feature details into that file.

Optional qualifiers:

-matchsource (string)
: By default, any feature source in the feature table is shown. You can set this to match any feature source you want to show. The source name is usually the name of the program that detected the feature, or the feature table (e.g., EMBL) that the feature came from. The source may be wildcarded by using *. If you want to show more than one source, separate their names with the | character. For example:

```
gene* | embl
```

-matchtype (string)
: By default, any feature type in the feature table is shown. You can set this to match any feature type you want to show. See Chapter 2 for a list of the EMBL feature types, and Chapter 3 for a list of the SWISS-PROT feature types. The type may be wildcarded using *. If you want to show more than one type, separate their names with the | character. For example:

```
*UTR | intron
```

-matchtag (string)
: Tags are the types of extra values that a feature may have. For example, in the EMBL feature table, a CDS type of feature may have the tags `/codon`, `/codon_start`, `/db_xref`, `/EC_number`, `/evidence`, `/exception`, `/function`, `/gene`, `/label`, `/map`, `/note`, `/number`, `/partial`, `/product`, `/protein_id`, `/pseudo`, `/standard_name`, `/translation`, `/transl_except`, `/transl_table`, or `/usedin`. Some of these tags also have values (e.g., `/gene` can have the value of the gene name). By default, any feature tag in the feature table is extracted. You can set this to match any feature tag you want to show. The tag may be wildcarded by using *. If you want to extract more than one tag, separate their names with the | character. For example:

```
gene | label
```

-matchvalue (string)

Tag values are the values associated with a feature tag. Tags are the types of extra values that a feature may have. For example, in the EMBL feature table, a CDS type of feature may have the tags /codon, /codon_start, /db_xref, /EC_number, /evidence, /exception, /function, /gene, /label, /map, /note, /number, /partial, /product, /protein_id, /pseudo, /standard_name, /translation, /transl_except, /transl_table, or /usedin. Some of these tags also have values (e.g., /gene can have the value of the gene name). By default, any feature tag in the feature table is extracted. You can set this to match any feature tag value you want to show. The tag may be wildcarded by using *. If you want to extract more than one tag, separate their names with the | character. For example:

```
pax* | 10
```

-sort (menu)

Sort features by Type, Start, or Source. Nosort uses the input order and does not sort. This qualifier can also be used to join coding regions together and leave other features in the input order.

Advanced qualifiers:

-html (boolean)

Use HTML formatting.

-[no]id (boolean)

Set this to false if you do not want to display the ID name of the sequence.

-[no]description (boolean)

Set this to false if you do not want to display the description of the sequence.

-[no]scale (boolean)

Set this to false if you do not want to display the scale line.

-width (integer)

You can expand (or contract) the width of the ASCII-character graphics display of the positions of the features using this value. For example, a width of 80 characters covers a standard page width, while a width a 10 characters is nearly unreadable. If the width is set to less than 4, the graphics lines and the scale line will not be displayed.

-collapse (boolean)

If set, features from the same source and of the same type and sense are all printed on the same line. For instance, if there are several features from the EMBL feature table (ie. the same source) which are all of the type "exon" in the same sense, they will all be displayed on the same line. This makes it hard to distinguish overlapping features. If this is set to false, each feature is displayed on a separate line. This makes it easier to distinguish where features start and end.

-[no]forward (boolean)

Set this to false if you do not want to display forward sense features.

-[no]reverse (boolean)

Set this to false if you do not want to display reverse sense features.

-[no]unknown (boolean)

Set this to false if you do not want to display unknown sense features. For example, features with no directionality (all protein features are of this type) and some nucleic features (such as CG-rich regions).

-strand (boolean)
: Set this if you want to display the strand of the features. Protein features are always directionless (indicated by `0`), forward is indicated by `+`, and reverse is `-`.

-source (boolean)
: Set this if you want to display the source of the features. The source name is usually either the name of the program that detected the feature or it is the name of the feature table (e.g., EMBL) that the feature came from.

-position (boolean)
: Set this if you want to display the start and end position of the features. If several features are being displayed on the same line, the start and end positions will be joined by a comma. For example: `189-189,225-225`.

-[no]type (boolean)
: Set this to false if you do not want to display the type of the features.

-tags (boolean)
: Set this to false if you do not want to display the tags and values of the features.

-[no]values (boolean)
: Set this to false if you do not want to display the tag values of the features. If this is set to false, only the tag names are displayed. If the tags are not displayed, the values will not be displayed. Because the value of the `translation` tag is often very long, it is never displayed.

showorf

showorf displays a nucleic acid sequence with its protein translation in a style suitable for publication. The translation can be done in any frame or combination of frames.

Here is a sample session with *showorf*:

```
% showorf
Pretty output of DNA translations
Input sequence: embl:paamir
Select Frames To Translate
         0 : None
         1 : F1
         2 : F2
         3 : F3
         4 : R1
         5 : R2
         6 : R3
Select one or more values [1,2,3,4,5,6]:
Output file [paamir.showorf]:
```

Mandatory qualifiers:

[-sequence] (sequence)
: Sequence USA.

-frames (menu)
: Select one or more values.

[-outfile] (outfile)
: Output filename.

Optional qualifiers:

-[no]ruler (boolean)
    Add a ruler.

-[no]plabel (boolean)
    Number translations.

-[no]nlabel (boolean)
    Number DNA sequence.

Advanced qualifiers:

-cfile (codon)
    Codon usage file.

-width (integer)
    Width of screen.

showseq

showseq displays a protein or a nucleic acid sequence in a style suitable for publication.

Here is a sample session with *showseq*. By default, the output appears on standard output (the terminal) but can be saved to a file. We only look at a small section of the sequence to save space:

```
% showseq tembl:eclac -sbeg 1 -send 100
Display a sequence with features, translation etc..
Things to display
         0 : Enter your own list of things to display
         1 : Sequence only
         2 : Default sequence with features
         3 : Pretty sequence
         4 : One frame translation
         5 : Three frame translations
         6 : Six frame translations
         7 : Restriction enzyme map
         8 : Baroque
Display format [2]:
Output file [eclac.showseq]: stdout
ECLAC
E.coli lactose operon with lacI, lacZ, lacY and lacA genes.

         10         20         30         40         50         60
----:----|----:----|----:----|----:----|----:----|----:----|
gacaccatcgaatggcgcaaaacctttcgcggtatggcatgatagcgcccggaagagagt
                  |
variation note="c in wild-type; t in 'up' promoter mutant I-Q [11]"
                                                 |=======
mRNA note="lacI (repressor) mRNA; preferred in vivo 3' end [12],[29]"

         70         80         90        100        110        120
----:----|----:----|----:----|----:----|----:----|----:----|
caattcagggtggtgaatgtgaaaccagtaacgttatacgatgtcgcagagtatgccggt
============================================================
mRNA note="lacI (repressor) mRNA; preferred in vivo 3' end [12],[29]"
                    |======================================
CDS codon_start="1" db_xref="SWISS-PROT:P03023" note="lac repressor p
```

Note that although we asked for the sequence display to end at position "100", it has displayed the sequence up to the end of the line - position "120". This is a feature of this program to make the display of things like restriction enzyme cutting sites easier.

The standard list of output formats are only a small selection of the possible ways in which a sequence might be displayed. Precise control over the output format is achieved by selecting the qualifier `-format 0` (Option 0 in the list of things to display). For example:

```
% showseq tembl:eclac -sbeg 1 -send 120
Display a sequence with features, translation etc..
Output file [stdout]:
Things to display
         0 : Enter your own list of things to display
         1 : Sequence only
         2 : Default sequence with features
         3 : Pretty sequence
         4 : One frame translation
         5 : Three frame translations
         6 : Six frame translations
         7 : Restriction enzyme map
         8 : Baroque
Display format [2]: 0
Specify your own things to display
         S : Sequence
         B : Blank line
         1 : Frame1 translation
         2 : Frame2 translation
         3 : Frame3 translation
        -1 : CompFrame1 translation
        -2 : CompFrame2 translation
        -3 : CompFrame3 translation
         T : Ticks line
         N : Number ticks line
         C : Complement sequence
         F : Features
         R : Restriction enzyme cut sites in forward sense
        -R : Restriction enzyme cut sites in reverse sense
         A : Annotation
Enter a list of things to display [B N T S A F]: b,s,t,c
Output file [eclac.showseq]: stdout
ECLAC
E.coli lactose operon with lacI, lacZ, lacY and lacA genes.

          gacaccatcgaatggcgcaaaacctttcgcggtatggcatgatagcgcccggaagagagt
          ----:----|----:----|----:----|----:----|----:----|----:----|
          ctgtggtagcttaccgcgttttggaaagcgccataccgtactatcgcgggccttctctca

          caattcagggtggtgaatgtgaaaccagtaacgttatacgatgtcgcagagtatgccggt
          ----:----|----:----|----:----|----:----|----:----|----:----|
          gttaagtcccaccacttacactttggtcattgcaatatgctacagcgtctcatacggcca
```

By choosing format "0" and specifying that we want to display the things: "b,s,t,c", we will output the sequence in the following way.

For every new line that the sequence starts to write, the output display will contain first a blank line ("b"), then the sequence itself ("s"), a line of with ticks every 10 characters ("t"), and the reverse complement of the sequence ("c'"). Subsequent lines of the sequence output will repeat this format.

The "thing" codes used in the list of standard formats are:

Sequence only:	S A
Default sequence:	B N T S A F
Pretty sequence:	B N T S A
One frame translation:	B N T S B 1 A F
Three frame translations:	B N T S B 1 2 3 A F
Six frame translations:	B N T S B 1 2 3 T -3 -2 -1 A F
Restriction enzyme map:	B R S N T C -R B 1 2 3 T -3 -2 -1 A
Baroque:	B 1 2 3 N T R S T C -R T -3 -2 -1 A F

The following are some examples of different formats.

Just sequence:

```
% showseq embl:eclac stdout -sbeg 1 -send 120 -noname -nodesc -format 0 -
thing S
Display a sequence with features, translation etc..
          gacaccatcgaatggcgcaaaacctttcgcggtatggcatgatagcgcccggaagagagt
          caattcagggtggtgaatgtgaaaccagtaacgttatacgatgtcgcagagtatgccggt
```

Protein sequence displayed in three-letter codes. The codes are displayed downwards, so the first code is "Met":

```
% showseq tsw:rs24_fugru stdout -three -format 2
RS24_FUGRU
40S RIBOSOMAL PROTEIN S24.

                   10        20        30        40        50        60
          ----:----|----:----|----:----|----:----|----:----|----:----|
          MAATVTVATALPMTAALLGALGMVVAVLHPGLATVPLTGIAGLLALMTLTTPAVVPVPGP
          esshaharhryhehsreelryleaasaeirlylharyhllrlyelyeyyhhrsaahahlh
          tnprlrlgrgsetrnguungsntllplusoysarlosruegusuastrsrroplleleye

                   70        80        90       100       110       120
          ----:----|----:----|----:----|----:----|----:----|----:----|
          ATGPGGGLTTGPAMVTASLATALLAGPLHALAAHGLPGLLLTSALGALGALAAMLLVAGT
          rhlhlllyhhlhleayseesylyyslryirelrilehlyyyherylrylrysreyyarlh
          grneyyysrryeatlrpruprassnuossguagsyueusssrrgsngsugsngtsslgyr

                  130       140       150       160       170       180
          ----:----|----:----|----:----|----:----|----:----|----:----|
          LLASVGASLLLA
          yylealleyyys
          ssarlyarsssp
```

Number the sequence lines in the margin:

```
% showseq tembl:mmam stdout -format 1 -number
Display a sequence with features, translation etc..
Output file [stdout]:
MMAM
Mus musculus (cell line C3H/F2-11) chromosome 12 anti-DNA antibody
heavy chain mRNA.
```

```
  1 gagnnccagctgcagcagtctggacctgagctggtaaagcctggggcttcagtgaagatg 60
 61 tcctgcaaggcttctggatacacattcactagctatgttatgcactgggtgaatcagaag 120
121 cctgggcagggccttgagtggattggatatattaatccttacaatgatggtactaactac 180
181 aatgagaagttcaaaggcaaggccacactgacttcagacaaatcctccagcacagcctac 240
241 atggagttcagcagcctgacctctgaggactctgcggtctattactgtgcaagaaaaact 300
301 tcctactatagtaacctatattactttgactactggggccaaggcaccactctcacagtc 360
361 tcctca                                                       366
```

Start the numbering at a specified value ("123" in this case):

```
% showseq tembl:mmam stdout -format 1 -number -offset 123
Display a sequence with features, translation etc..
MMAM
Mus musculus (cell line C3H/F2-11) chromosome 12 anti-DNA antibody
heavy chain mRNA.
      123 gagnnccagctgcagcagtctggacctgagctggtaaagcctggggcttcagtgaagatg 182
      183 tcctgcaaggcttctggatacacattcactagctatgttatgcactgggtgaatcagaag 242
      243 cctgggcagggccttgagtggattggatatattaatccttacaatgatggtactaactac 302
      303 aatgagaagttcaaaggcaaggccacactgacttcagacaaatcctccagcacagcctac 362
      363 atggagttcagcagcctgacctctgaggactctgcggtctattactgtgcaagaaaaact 422
      423 tcctactatagtaacctatattactttgactactggggccaaggcaccactctcacagtc 482
      483 tcctca                                                       488
```

Make selected regions uppercase. Use -slower to force the rest of the sequence to be lowercase:

```
% showseq tembl:mmam stdout -format 1 -slower -upper '25-45,101-203,333-362'
Display a sequence with features, translation etc..
MMAM
Mus musculus (cell line C3H/F2-11) chromosome 12 anti-DNA antibody
heavy chain mRNA.
          gagnnccagctgcagcagtctggaCCTGAGCTGGTAAAGCCTGGGgcttcagtgaagatg
          tcctgcaaggcttctggatacacattcactagctatgttaTGCACTGGGTGAATCAGAAG
          CCTGGGCAGGGCCTTGAGTGGATTGGATATATTAATCCTTACAATGATGGTACTAACTAC
          AATGAGAAGTTCAAAGGCAAGGCcacactgacttcagacaaatcctccagcacagcctac
          atggagttcagcagcctgacctctgaggactctgcggtctattactgtgcaagaaaaact
          tcctactatagtaacctatattactttgactaCTGGGGCCAAGGCACCACTCTCACAGTC
          TCctca
```

Translate selected regions:

```
% showseq embl:mmam tstdout -format 4 -send 120 -trans 25-49,66-76
Display a sequence with features, translation etc..
MMAM
Mus musculus (cell line C3H/F2-11) chromosome 12 anti-DNA antibody
heavy chain mRNA.

                 10        20        30        40        50        60
          ----:----|----:----|----:----|----:----|----:----|----:----|
          gagnnccagctgcagcagtctggacctgagctggtaaagcctggggcttcagtgaagatg

                                  P  E  L  V  K  P  G  A  S

                 70        80        90       100       110       120
          ----:----|----:----|----:----|----:----|----:----|----:----|
          tcctgcaaggcttctggatacacattcactagctatgttatgcactgggtgaatcagaag

                R  L  L
```

Add your own annotation to the display:

```
% showseq tembl:mmam stdout -format 2 -send 120 -annotation '13-26 binding
site 15-15 SNP'
Display a sequence with features, translation etc..
MMAM
Mus musculus (cell line C3H/F2-11) chromosome 12 anti-DNA antibody
heavy chain mRNA.

                  10        20        30        40        50        60
          ----:----|----:----|----:----|----:----|----:----|----:----|
          gagnnccagctgcagcagtctggacctgagctggtaaagcctggggcttcagtgaagatg
                      |------------|
                      binding site
                        |
                        SNP

                  70        80        90       100       110       120
          ----:----|----:----|----:----|----:----|----:----|----:----|
          tcctgcaaggcttctggatacacattcactagctatgttatgcactgggtgaatcagaag
```

Mandatory qualifiers (bold if not always prompted):

[-sequence] (seqall)
Sequence database USA.

-format (menu)
Display format.

-things *(menu)*
Specify a list of one or more code characters in the order in which you want things to be displayed. If you want to see things displayed in the order: sequence, complement sequence, ticks line, frame 1 translation, and blank line, enter: `S,C,T,1,B`.

[-outfile] (outfile)
If you enter the name of a file here, this program will write the sequence details into that file.

Optional qualifiers:

-translate (range)
Regions to translate (if translating). If this is left blank the complete sequence is translated. A set of regions is specified by a set of pairs of positions. The positions are integers. They are separated by any non-digit, non-alpha character. Examples of region specifications are:

```
24-45, 56-78
1:45, 67=99;765..888
1,5,8,10,23,45,57,99
```

-uppercase (range)
Regions to put in uppercase. If this is left blank, the sequence case is left alone. A set of regions is specified by a set of pairs of positions. The positions are integers. They are separated by any non-digit, non-alpha character. Examples of region specifications are:

```
24-45, 56-78
1:45, 67=99;765..888
1,5,8,10,23,45,57,99
```

-highlight (range)

Regions to color if formatting for HTML. If this is left blank, the sequence is left alone. A set of regions is specified by a set of pairs of positions. The positions are integers. They are followed by any valid HTML font color. Examples of region specifications are:

```
24-45 blue 56-78 orange
1-100 green 120-156 red
```

A file of ranges to color (one range per line) can be specified as `@filename`.

-annotation (range)

Regions to annotate by marking. If this is left blank, no annotation is added. A set of regions is specified by a set of pairs of positions followed by optional text. The positions are integers. They are followed by any text (but not digits when on the command-line). Examples of region specifications are:

```
24-45 new domain 56-78 match to Mouse
1-100 First part 120-156 oligo
```

A file of ranges to annotate (one range per line) can be specified as `@filename`.

-enzymes (string)

The argument `all` reads all enzyme names from the REBASE database. You can specify enzymes by giving their names with commas between them, such as: `HincII,hinfI,ppiI,hindiii`. This command is not case-sensitive. You may also use the data from file containing enzyme names by prepending the name of the file you want to use with an `@` character; for example, `@enz.list`. Blank lines and lines starting with a comment tag (# or !) within the file are ignored; all other lines are concatenated together with a comma and treated as the list of enzymes to search for. A file containing enzyme names might look like this:

```
! my enzymes
HincII, ppiII
! other enzymes
hindiii
HinfI
PpiI
```

-table (menu)

Code to use. See the *fuzztran* description for codes.

-matchsource (string)

By default, any feature source in the feature table is shown. You can set this to match any feature source you want to show. The source name is usually the name of the program that detected the feature, or the feature table (e.g., EMBL) that the feature came from. The source may be wildcarded by using *. If you want to show more than one source, separate their names with the | character. For example:

```
gene* | embl
```

-matchtype (string)

By default, any feature type in the feature table is shown. You can set this to match any feature type you want to show. See Chapter 2 for a list of the EMBL feature types, and Chapter 3 for a list of the SWISS-PROT feature types. The type may be wildcarded by using the * character. If you want to show more than one type, separate their names with the | character. For example:

```
*UTR | intron
```

-matchsense (integer)
By default, any feature type in the feature table is shown. You can set this to match any feature sense you want to show. `0` = any sense, `1` = forward sense, and `-1` = reverse sense.

-minscore (float)
If this is greater than or equal to the maximum score, any score is permitted.

-maxscore (float)
If this is less than or equal to the maximum score, any score is permitted.

-matchtag (string)
Tags are the types of extra values that a feature may have. For example, in the EMBL feature table, a CDS type of feature may have the tags `/codon`, `/codon_start`, `/db_xref`, `/EC_number`, `/evidence`, `/exception`, `/function`, `/gene`, `/label`, `/map`, `/note`, `/number`, `/partial`, `/product`, `/protein_id`, `/pseudo`, `/standard_name`, `/translation`, `/transl_except`, `/transl_table`, or `/usedin`. Some of these tags also have values (e.g., `/gene` can have the value of the gene name). By default, any feature tag in the feature table is extracted. You can set this to match any feature tag you want to show. The tag may be wildcarded by using *. If you want to extract more than one tag, separate their names with the | character. For example:

```
gene | label
```

-matchvalue (string)
Tag values are the values associated with a feature tag. Tags are the types of extra values that a feature may have. For example, in the EMBL feature table, a CDS type of feature may have the tags `/codon`, `/codon_start`, `/db_xref`, `/EC_number`, `/evidence`, `/exception`, `/function`, `/gene`, `/label`, `/map`, `/note`, `/number`, `/partial`, `/product`, `/protein_id`, `/pseudo`, `/standard_name`, `/translation`, `/transl_except`, `/transl_table`, or `/usedin`. Some of these tags also have values (e.g., `/gene` can have the value of the gene name). By default, any feature tag in the feature table is extracted. You can set this to match any feature tag value you want to show. The tag may be wildcarded by using *. If you want to extract more than one tag, separate their names with the | character. For example:

```
pax* | 10
```

Advanced qualifiers:

-orfminsize (integer)
Minimum size of Open Reading Frames (ORFs) to display in the translations.

-flatreformat (boolean)
Display restriction enzyme sites in flat format.

-mincuts (integer)
Minimum cuts per restriction enzyme.

-maxcuts (integer)
Maximum cuts per restriction enzyme.

-sitelen (integer)
Minimum recognition site length.

-single (boolean)
Force single restriction enzyme site only cuts.

-[no]blunt (boolean)
Allow blunt end restriction enzyme cutters.

-[no]sticky (boolean)
Allow sticky end restriction enzyme cutters.

-[no]ambiguity (boolean)
Allow ambiguous restriction enzyme matches.

-plasmid (boolean)
Allow circular DNA.

-[no]commercial (boolean)
Only use restriction enzymes with suppliers.

-[no]limit (boolean)
Limits restriction enzyme hits to one isoschizomer.

-preferred (boolean)
Report preferred isoschizomers.

-threeletter (boolean)
Display protein sequences in three-letter code.

-number (boolean)
Number the sequences.

-width (integer)
Width of sequence to display.

-length (integer)
Line length of page (0 for indefinite).

-margin (integer)
Margin around sequence for numbering.

-[no]name (boolean)
Set this to false if you do not want to display the ID name of the sequence.

-[no]description (boolean)
Set this to false if you do not want to display the description of the sequence.

-offset (integer)
Offset to start numbering the sequence from.

-html (boolean)
Use HTML formatting.

shuffleseq

shuffleseq takes a sequence as input and outputs one or more sequences whose order has been randomly shuffled. No bases or residues are changed, only their order.

Here is a sample session with *shuffleseq* making two randomised copies of the input sequence:

```
% shuffleseq -shuffle 2
Shuffles a set of sequences maintaining composition
Input sequence(s): embl:mmam
Output sequence [mmam.fasta]:
```

Mandatory qualifiers:

[-sequence] (seqall)
Sequence database USA.

[-outseq] (seqoutall)
Output sequence(s) USA.

EMBOSS

Advanced qualifiers:

-shuffle (integer)
: Number of shuffles.

sigcleave

sigcleave reports protein signal cleavage sites.

Here is a sample session with *sigcleave*:

```
% sigcleave
Reports peptide signal cleavage sites
Input sequence: sw:ach2_drome
Output file [ach2_drome.out]:
Minimum weight [3.5]:
```

Mandatory qualifiers:

[-sequence] (seqall)
: Sequence database USA.

-minweight (float)
: Minimum scoring weight value for the predicted cleavage site.

[-outfile] (report)
: Output report filename.

Optional qualifiers:

-prokaryote (boolean)
: Specifies the sequence is prokaryotic and changes the default scoring data filename.

Advanced qualifiers:

-pval (integer)
: Specifies the number of columns before the residue at the cleavage site in the weight matrix table.

-nval (integer)
: Specifies the number of columns after the residue at the cleavage site in the weight matrix table.

siggen

siggen parses a multiple structure alignment generated by the EMBOSS application *scopalign* and corresponding files of residue contact data generated by the EMBOSS application contacts and generates a protein signature of a specified sparsity.

Here is a sample session with *siggen*:

```
% siggen
Generates a sparse protein signature
Location of alignment files for input [./]: ./jontest
Extension of alignment files for input [.align]:
Location of contact files for input [./]: ./jontest
Extension of contact files [.con]:
```

```
% sparsity of signature [10]:
Generate a randomized signature [N]:
Substitution matrix to be used [./EBLOSUM62]:
Score alignment on basis of residue conservation [Y]:
Score alignment on basis of number of contacts [Y]:
Score alignment on basis of conservation of contacts [Y]: N
Score alignment on a combined measure of number and conservation of contacts
[N]:
Ignore alignment postitions with post_similar value of 0 [Y]:
Name of signature file for output [sig.sig]:
```

Mandatory qualifiers (bold if not always prompted):

[-algpath] (string)
: Location of scop structure-based sequence alignment files (input).

[-algextn] (string)
: Extension of alignment files.

-sparsity (integer)
: Percentage sparsity of signature.

-seqoption *(menu)*
: Select number.

-datafile *(matrixf)*
: This is the scoring matrix file used when comparing sequences.

-conoption *(menu)*
: Select number.

-filtercon *(boolean)*
: Ignore alignment positions making less than a threshold number of contacts.

-conthresh *(integer)*
: Threshold contact number.

-conpath *(string)*
: Location of contact files (input).

-conextn *(string)*
: Extension of contact files.

-cpdbpath *(string)*
: Location of domain coordinate files (EMBL format input).

-cpdbextn *(string)*
: Extension of coordinate files.

-filterpsim *(boolean)*
: Ignore alignment postitions with post_similar value of 0.

[-sigpath] (string)
: Location of signature files (output).

[-sigextn] (string)
: Extension of signature files.

Advanced qualifiers:

-randomise (boolean)
: Generate a randomized signature.

sigscan

sigscan scans a signature such as that generated by the EMBOSS application *siggen* against a protein sequence database and generates files of scored hits and corresponding alignments.

Here is a sample session with *sigscan*:

```
% sigscan
```

Mandatory qualifiers:

[-sigin] (infile)
: Name of signature file (input).

-database (seqall)
: Name of the SWISS-PROT sequence database to search.

-targetf (infile)
: Name of validation (input).

-thresh (integer)
: Minimum length (residues) of overlap required for two hits with the same code to count as the same hit.

-sub (matrixf)
: This is the scoring matrix file used when comparing sequences.

-gapo (float)
: The gap insertion penalty is the score taken away when a gap is created. The best value depends on the choice of comparison matrix. The default value assumes you are using the EBLOSUM62 matrix for protein sequences, and the EDNAMAT matrix for nucleotide sequences.

-gape (float)
: The gap extension penalty is added to the standard gap penalty for each base or residue in the gap. This is how long gaps are penalized. You can usually expect a few long gaps rather than many short gaps, so the gap extension penalty should be lower than the gap penalty. An exception occurs when one or both sequences are single reads with possible sequencing errors, in which case you should expect many single base gaps. You can obtain this result by setting the gap open penalty to zero (or a very low value) and using the gap extension penalty to control gap scoring.

-nterm (menu)
: Select number.

-nhits (integer)
: Number of hits to output.

[-hitsf] (outfile)
: Name of signature hits file (output).

[-alignf] (outfile)
: Name of signature alignments file (output).

silent

silent does a scan of a nucleic acid sequence for silent mutation restriction enzyme sites.

Here is a sample session with *silent*:

```
% silent
Silent mutation restriction enzyme scan
Input sequence: embl:hsfau
Comma separated enzyme list [all]: ecori,hindiii
Output file [hsfau.silent]:
```

Mandatory qualifiers:

[-seq] (sequence)
  Sequence USA.

-enzymes (string)
  Comma-separated enzyme list.

[-outf] (outfile)
  Output filename.

Advanced qualifiers:

-sshow (boolean)
  Display untranslated sequence.

-tshow (boolean)
  Display translated sequence.

-allmut (boolean)
  Display all mutations.

skipseq

skipseq skips the first few sequences in a multiple set of sequences, and writes out the rest of them. *skipseq* is a variant of the standard program for reading and writing sequences, *seqret*.

Here is a sample session with *skipseq*:

```
% skipseq -skip 1
Reads and writes (returns) sequences, skipping the first few
Input sequence(s): tembl:eclac*
Output sequence [eclac.fasta]:
```

Mandatory qualifiers:

[-sequence] (seqall)
  Sequence database USA.

-skip (integer)
  Number of sequences to skip at start.

[-outseq] (seqoutall)
  Output sequence(s) USA.

Advanced qualifiers:

-feature (boolean)
  Use feature information.

splitter

splitter is a simple editing program that allows you to split a long sequence into smaller, optionally overlapping, subsequences.

To split a sequence into subsequences of 10,000 bases (the default size) with no overlap between the subsequences:

```
% splitter one_huge.seq many_small.seq
```

To split a sequence into subsequences of 50,000 bases with an overlap of 3,000 bases on each subsequence:

```
% splitter one_huge.seq many_small.seq -size=50000 -over=3000
```

Mandatory qualifiers:

[-sequence] (seqall)
  Sequence database USA.

[-outseq] (seqoutall)
  Output sequence(s) USA.

Optional qualifiers:

-size (integer)
  Size to split at.

-overlap (integer)
  Overlap between split sequences.

stretcher

stretcher calculates a global alignment of two sequences using a modification of the classic dynamic programming algorithm which uses linear space.

Here is a sample session with *stretcher*:

```
% stretcher tsw:hba_human tsw:hbb_human
Finds the best global alignment between two sequences
Output alignment [hba_human.stretcher]:
```

Mandatory qualifiers:

[-sequencea] (sequence)
  Sequence USA.

[-sequenceb] (sequence)
  Sequence USA.

[-outfile] (align)
  Output alignment filename.

Optional qualifiers:

-datafile (matrix)
: This is the scoring matrix file used when comparing sequences. By default, it is the file *EBLOSUM62* (for proteins) or the file *EDNAFULL* (for nucleic sequences). These files are found in the *data* directory of the EMBOSS installation.

-gappenalty (integer)
: Gap penalty.

-gaplength (integer)
: Gap length penalty.

---

stssearch

stssearch searches a DNA sequence database with a set of STS primers and reports expected matches and possible mismatches.

Here is a sample session with *stssearch*:

```
% stssearch
Input sequence(s): embl:eclac*
Primer file: lac.primers
```

Mandatory qualifiers:

[-sequences] (seqall)
: Sequence database USA.

[-primers] (infile)
: Primer file.

[-out] (outfile)
: Output filename.

---

supermatcher

supermatcher finds a match of a large sequence against one or more sequences.

Here is a sample session with *supermatcher*:

```
% supermatcher tembl:ec\* tembl:eclac -word 50 -sbegin2 101 -send2 -101
Finds a match of a large sequence against one or more sequences
Gap opening penalty [10.0]: 3.0
Gap extension penalty [0.5]:
Output alignment [eclac.supermatcher]:
```

Mandatory qualifiers:

[-seqa] (seqall)
: Sequence database USA.

[-seqb] (seqset)
: Sequence set USA.

-gapopen (float)
: Gap opening penalty.

-gapextend (float)
: Gap extension penalty.

-outfile (align)
: Output alignment filename.

Optional qualifiers:

-datafile (matrixf)
: This is the scoring matrix file used when comparing sequences. By default, it is the file *EBLOSUM62* (for proteins) or the file *EDNAFULL* (for nucleic sequences). These files are found in the *data* directory of the EMBOSS installation.

-width (integer)
: Alignment width.

-wordlen (integer)
: word length for initial matching.

-errorfile (outfile)
: Error file to write to.

---

swissparse

swissparse retrieves sequences from SWISS-PROT using keyword search.

Here is a sample session with *swissparse*:

```
% swissparse
```

Mandatory qualifiers:

[-keyfile] (infile)
: Name of keywords file for input.

[-spfile] (infile)
: Name of SWISS-PROT database to read.

[-outfile] (outfile)
: Name of search results file for output.

---

syco

syco is a frame-specific gene finder that tries to recognize protein coding sequences by virtue of the similarity of their codon usage to a codon frequency table.

Here is a sample session with *syco*, requesting graphical output:

```
% syco -plot -cfile pae
Synonymous codon usage Gribskov statistic plot
Input sequence: embl:paamir
Graph type [x11]:
```

It is essential to use the correct codon usage file for the species.

Mandatory qualifiers (bold if not always prompted):

[-sequence] (sequence)
: Sequence USA.

-graph *(xygraph)*
Graph type.

-outfile *(outfile)*
Output filename.

Advanced qualifiers:

-cfile (codon)
Codon usage file.

-window (integer)
Averaging window.

-uncommon (boolean)
Show common codon usage.

-minimum (float)
Minimum value for a common codon.

-plot (boolean)
Produce plot.

EMBOSS

---

textsearch

textsearch is a small utility search for words in the description text of a sequence and for each match list the sequence's name and/or description. It only searches the description line of the annotation, not the full annotation.

Search for matches to "lacZ":

```
% textsearch swissprot:\*  'lacz' stdout
```

Search for matches to "lacz" or "permease" in E.coli proteins:

```
% textsearch swissprot:\*_ecoli 'lacZ | permease' stdout
```

Output a search for "transport" formatted with HTML to a file:

```
% textsearch embl:\* 'transport' -html -outfile embl.transport stdout
```

Mandatory qualifiers:

[-sequence] (seqall)
Sequence database USA.

[-pattern] (string)
The search pattern is a regular expression. Use a | to indicate OR. For example: human|mouse will find text with either "human" OR "mouse" in the text

[-outfile] (outfile)
If you enter the name of a file here, this program will write the sequence details into that file.

Optional qualifiers:

-casesensitive (boolean)
Do a case-sensitive search.

-html (boolean)
Format output as an HTML table.

Advanced qualifiers:

-only (boolean)
: This is a way of shortening the command line if you only want to display a few things. Instead of specifying: `-nohead -noname -nousa -noacc -nodesc` to get only the name output, you can specify `-only -name`.

-heading (boolean)
: Display column headings.

-usa (boolean)
: Display the USA of the sequence.

-accession (boolean)
: Display "accession" column.

-name (boolean)
: Display "name" column.

-description (boolean)
: Display "description" column.

---

tfextract

tfextract extracts data from TRANSFAC.

Here is a sample session with *tfextract*:

```
% tfextract
Extract data from TRANSFAC
Full pathname of transfac SITE.DAT: /data/transfac/site.dat
```

Mandatory qualifiers:

[-inf] (infile)
: Full pathname of transfac SITE.DAT.

---

tfm

tfm displays the help documentation for an EMBOSS program.

Here is a sample session with *tfm*:

```
% tfm wossname
```

You will see a page of documentation on the *wossname* program. To see further pages, you should press the spacebar on your keyboard. To stop seeing the documentation before you get to the end of the text, you can press the "q" key to quit.

Mandatory qualifiers:

[-program] (string)
: Enter the name of an EMBOSS program.

Optional qualifiers:

-outfile (outfile)
: If you enter the name of a file here, this program will write the program names and brief descriptions into that file.

-html (boolean)
: This will format the output for displaying as a WWW document.

-more (boolean)
This uses the standard Unix utility *more* to display the text page-by-page, waiting for you to read one screen of text before going on to the next page. When you have finished reading a page, press the spacebar to proceed to the next page.

tfscan

tfscan takes a sequence and the name of one of these taxonomic groups and does a fast match of the TRANSFAC sequences against the input sequence (optionally allowing mismatches).

Here is a sample session with *tfscan*:

```
% tfscan
Input sequence(s): embl:hsfos
Transcription Factor Class
         F : fungi
         I : insect
         P : plant
         V : vertebrate
         O : other
Select class [V]: v
Number of mismatches [0]:
Output file [hsfos.tfscan]:
```

Mandatory qualifiers:

[-sequence] (seqall)
Sequence database USA.

-menu (menu)
Select class.

-mismatch (integer)
Number of mismatches.

[-outfile] (outfile)
Output filename.

tmap

tmap predicts transmembrane segments in proteins.

Here is a sample session with *tmap*:

```
% tmap sw:opsd_human -out tmap.res
```

Mandatory qualifiers:

[-msf] (seqset)
File containing a sequence alignment.

-graph (xygraph)
Graph type.

Optional qualifiers:

-outfile (outfile)
Output filename.

tranalign

tranalign is a simple program that allows you to produce aligned cDNA sequences from aligned protein sequences.

Here is a sample session with *tranalign*:

```
% tranalign tranalign.seq tranalign.pep tranalign2.seq
```

Mandatory qualifiers:

[-nsequence] (seqall)
Nucleotide sequences to align.

[-psequence] (seqset)
Protein sequence alignment.

[-outseq] (seqoutset)
Output sequence set USA.

Optional qualifiers:

-table (menu)
Code to use. See the *fuzztran* description for codes.

transeq

transeq translates nucleic acid sequences to the corresponding peptide sequence.

To translate a sequence "pop.seq" in the first frame (starting at the first base and proceeding to the end):

```
% transeq pop.seq pop.pep
```

To translate a sequence "pop.seq" in the second frame:

```
% transeq pop.seq pop.pep -frame=2
```

To translate a sequence "pop.seq" in the third frame in the reverse sense (starting at the last base and proceeding to the start):

```
% transeq pop.seq pop.pep -frame=-1
```

To translate a sequence "pop.seq" in all three forward frames:

```
% transeq pop.seq pop.pep -frame=F
```

To translate a sequence "pop.seq" in all three reverse frames:

```
% transeq pop.seq pop.pep -frame=R
```

To translate a sequence "pop.seq" in all six forward and reverse frames:

```
% transeq pop.seq pop.pep -frame=6
```

To translate a specific set of regions corresponding to a known set of coding sequences:

```
% transeq pop.seq pop.pep -reg=2-45,67-201,328-509
```

To translate a sequence "mito.seq" using the mammalian mitochondrion genetic code table:

```
% transeq mito.seq mito.pep -table=2
```

Mandatory qualifiers:

[-sequence] (seqall)
Sequence database USA.

[-outseq] (seqoutall)
Output sequence(s) USA.

Optional qualifiers:

-frame (menu)
Frame(s) to translate.

-table (menu)
Code to use. See the *fuzztran* description for codes.

-regions (range)
Regions to translate. If this is left blank, the complete sequence is translated. A set of regions is specified by a set of pairs of positions. The positions are integers. They are separated by any non-digit, non-alpha character. Examples of region specifications are:

```
24-45, 56-78
1:45, 67=99;765..888
1,5,8,10,23,45,57,99
```

Note: you should not try to use this option with any other frame than the default, `-frame=1`.

-trim (boolean)
This removes all X and asterisk characters from the right end of the translation. The trimming process starts at the end and continues until the next character is not an X or an asterisk.

Advanced qualifiers:

-alternative (boolean)
The default definition of `frame -1` is the reverse-complement of the set of codons used in frame 1. (`frame -2` is the set of codons used by frame 2, similarly `frames -3` is the set used by 3). This is a common standard, used by the Staden package and other programs. If you prefer to define `frame -1` as using the set of codons starting with the last codon of the sequence, set this to true.

trimest

trimest trims poly-A tails off EST sequences.

Here is a sample session with *trimest*:

```
% trimest embl:hsfau hsfau.seq
```

Mandatory qualifiers:

[-sequence] (seqall)
Sequence database USA.

[-outseq] (seqoutall)
Output sequence(s) USA.

Optional qualifiers:

-minlength (integer)

This is the minimum length that a poly-A (or poly-T) tail must have before it is removed. If there are mismatches in the tail, there must be at least this length of poly-A tail before the mismatch for the mismatch in order to be considered as part of the tail.

-mismatches (integer)

If there are this number or fewer contiguous non-A bases in a poly-A tail and there are `-minlength A` bases before them, they are considered part of the tail and removed. For example, the terminal 4 A's of GCAGAAAA would be removed with the default values of `-minlength=4` and `-mismatches=1` (There are not at least 4 A's before the last "G," so only the A's after it are considered to be part of the tail). The terminal 9 bases of GCAAAAGAAAA would be removed; there are at least `-minlength A`'s preceeding the last "G", so it is part of the tail.

-[no]reverse (boolean)

When a poly-T region at the 5' end of the sequence is found and removed, it is likely that the sequence is in the reverse sense. This option will change the sequence to the forward sense when it is written out. If this option is not set, the sense will not be changed.

Advanced qualifiers:

-[no]fiveprime (boolean)

If this is set true, the 5' end of the sequence is inspected for poly-T tails. These are removed if they are longer than any 3' poly-A tails. If this is false, the 5' end is ignored.

trimseq

trimseq is used to tidy up the ends of sequences, removing all the bits that you would really rather were not published.

Tidy up the sequence ends, stopping at the first wanted code:

```
% trimseq xyz.seq xyz_clean.seq -window 1 -percent 100
```

Tidy up the sequence ends, removing poor bits at the ends:

```
% trimseq xyz.seq xyz_clean.seq -window 5 -percent 40
```

Tidy up the sequence ends, removing very poor bits at the ends:

```
% trimseq xyz.seq xyz_clean.seq -window 20 -percent 80
```

Tidy up the sequence ends, removing even maginally poor bits at the ends:

```
% trimseq xyz.seq xyz_clean.seq -window 20 -percent 10
```

Tidy up the sequence ends, removing poor bits including ambiguity codes:

```
% trimseq xyz.seq xyz_clean.seq -window 20 -percent 50 -strict
```

Tidy up the sequence ends, removing asterisks from a protein end:

```
% trimseq xyz.seq xyz_clean.seq -window 1 -percent 100 -star
```

Tidy up the sequence ends, removing poor bits at only the left end:

```
% trimseq xyz.seq xyz_clean.seq -window 20 -percent 50 -noright
```

Mandatory qualifiers:

[-sequence] (seqall)
: Sequence database USA.

[-outseq] (seqoutall)
: Output sequence(s) USA.

Optional qualifiers:

-window (integer)
: This determines the size of the region that is considered when deciding whether the percentage of ambiguity is greater than the threshold. A value of 5 means that a region of 5 letters in the sequence is shifted along the sequence from the ends and trimming is done only if there is a greater or equal percentage of ambiguity than the threshold percentage.

-percent (float)
: This is the threshold of the percentage ambiguity in the window required in order to trim a sequence.

-strict (boolean)
: In nucleic sequences, trim off not only Ns and Xs, but also the nucleotide IUPAC ambiguity codes M, R, W, S, Y, K, V, H, D and B. In protein sequences, trim off not only Xs but also B and Z.

-star (boolean)
: In protein sequences, trim off not only Xs, but the asterisks as well.

Advanced qualifiers:

-[no]left (boolean)
: Trim at the start.

-[no]right (boolean)
: Trim at the end.

union

union reads in several sequences, concatenates them, and writes them out as a single sequence.

Here is a sample session with *union*. The file *cds.list* contains a list of the regions making up the coding sequence of the sequence "embl:hsfau":

```
% union
Reads sequence fragments and builds one sequence
Input sequence(s): @cds.list
Output sequence [hsfau1.fasta]: fau.cds
```

Mandatory qualifiers:

[-sequence] (seqall)
: Sequence database USA.

[-outseq] (seqout)
: Output sequence USA.

vectorstrip

vectorstrip is useful for stripping vector sequence from the ends of sequences of interest.

Here are several examples of running *vectorstrip*. The vectorfile and the sequence files used in these example are given below. In each case, the same fragment has been cloned into the XhoI site of the polylinker of each vector. The cloned fragment is represented in lowercase, and the vector sequence is represented in uppercase. This makes it easy to see the sequence trimming.

1. Running *vectorstrip* on a list of sequences with default parameters:

```
% vectorstrip @seqs.list
Strips out DNA between a pair of vector sequences
Are your vector sequences in a file? [Y]:
Name of vectorfile: vectors
Max allowed % mismatch [10]:
Show only the best hits (minimise mismatches)? [Y]:
Output file [pbluescript.vectorstrip]:stdout
Output sequence [pbluescript.fasta]:

Sequence: pBlueScript    Vector: pTYB1  No match

Sequence: pBlueScript    Vector: pBS_KS+
5' sequence matches:
        From 67 to 83 with 0 mismatches
3' sequence matches:
        From 205 to 219 with 0 mismatches
Sequences output to file:
        from 84 to 204
                tcgagagccgtattgcgatatagcgcacatgcgttggacacagatgagca
                cacagtgacatgagagacacagatatagagacagatagacgatagacaga
                cagcatatatagacagatagc
        sequence trimmed from 5' end:
                GGAAACAGCTAATGACCATGATTACGCCAAGCGCGCAATTAACCCTCACT
                AAAGGGAACAAAAGCTGGGTACCGGGCCCCCCC
        sequence trimmed from 3' end:
                TCGAGGTCGACGGTATCGATAAGCTTGATATCG

Sequence: pBlueScript    Vector: pLITMUS        No match

Sequence: litmus.seq     Vector: pTYB1  No match

Sequence: litmus.seq     Vector: pBS_KS+        No match

Sequence: litmus.seq     Vector: pLITMUS
5' sequence matches:
        From 43 to 61 with 0 mismatches
3' sequence matches:
        From 183 to 199 with 0 mismatches
Sequences output to file:
        from 62 to 182
                tcgagagccgtattgcgatatagcgcacatgcgttggacacagatgagca
                cacagtgacatgagagacacagatatagagacagatagacgatagacaga
```

```
                cagcatatatagacagatagc
        sequence trimmed from 5' end:
                TCTAGAACCGGTGACGTCTCCCATGGTGAAGCTTGGATCCACGATATCCT
                GCAGGAATTCC

Sequence: pTYB1.seq      Vector: pTYB1
5' sequence matches:
        From 40 to 58 with 0 mismatches
3' sequence matches:
        From 180 to 196 with 0 mismatches
Sequences output to file:
        from 59 to 179
                tcgagagccgtattgcgatatagcgcacatgcgttggacacagatgagca
                cacagtgacatgagagacacagatatagagacagatagacgatagacaga
                cagcatatatagacagatagc
        sequence trimmed from 5' end:
                CTTTAAGAAGGAGATATACATATGGCTAGCTCGCGAGTCGACGGCGGCCG
                CGAATTCC
        sequence trimmed from 3' end:
                TCGAGGGCTCTTCCTGCTTTGCCAAGGGTACCAATGTTTTAATGGCGGAT

Sequence: pTYB1.seq      Vector: pBS_KS+        No match

Sequence: pTYB1.seq      Vector: pLITMUS        No match

% more pbluescript.fasta
>pBlueScript_from_84_to_204 KS+
tcgagagccgtattgcgatatagcgcacatgcgttggacacagatgagcacacagtgaca
tgagagacacagatatagagacagatagacgatagacagacagcatatatagacagatag
c
>litmus.seq_from_62_to_182
tcgagagccgtattgcgatatagcgcacatgcgttggacacagatgagcacacagtgaca
tgagagacacagatatagagacagatagacgatagacagacagcatatatagacagatag
c
>pTYB1.seq_from_59_to_179
tcgagagccgtattgcgatatagcgcacatgcgttggacacagatgagcacacagtgaca
tgagagacacagatatagagacagatagacgatagacagacagcatatatagacagatag
c
```

2. Running *vectorstrip* allowing maximum 30 percent mismatch, then asking only for best hits:

```
% vectorstrip litmus.seq
Strips out DNA between a pair of vector sequences
Are your vector sequences in a file? [Y]:
Name of vectorfile: vectors
Max allowed % mismatch [10]: 30
Show only the best hits (minimise mismatches)? [Y]:
Output file [litmus.vectorstrip]: stdout
Output sequence [litmus.fasta]:

Sequence: litmus.seq     Vector: pTYB1  No match

Sequence: litmus.seq     Vector: pBS_KS+        No match
```

```
Sequence: litmus.seq     Vector: pLITMUS
5' sequence matches:
        From 43 to 61 with 0 mismatches
3' sequence matches:
        From 183 to 199 with 0 mismatches
Sequences output to file:
        from 62 to 182
                tcgagagccgtattgcgatatagcgcacatgcgttggacacagatgagca
                cacagtgacatgagagacacagatatagagacagatagacgatagacaga
                cagcatatatagacagatagc
        sequence trimmed from 5' end:
                TCTAGAACCGGTGACGTCTCCCATGGTGAAGCTTGGATCCACGATATCCT
                GCAGGAATTCC
        sequence trimmed from 3' end:
                TCGAGACCGTACGTGCGCGCGAATGCATCCAGATCTTCCCTCTAGTCAAG
                GCCTTAAGTGAGTCGTATTACGGA
```

3. Running *vectorstrip* allowing maximum 30 percent mismatch, then asking for all hits:

```
% vectorstrip litmus.seq
Strips out DNA between a pair of vector sequences
Are your vector sequences in a file? [Y]:
Name of vectorfile: vectors
Max allowed % mismatch [10]: 30
Show only the best hits (minimise mismatches)? [Y]: N
Output file [litmus.vectorstrip]: stdout
Output sequence [litmus.fasta]:

Sequence: litmus.seq     Vector: pTYB1  No match

Sequence: litmus.seq     Vector: pBS_KS+        No match

Sequence: litmus.seq     Vector: pLITMUS
5' sequence matches:
        From 43 to 61 with 0 mismatches
3' sequence matches:
        From 183 to 199 with 0 mismatches
        From 228 to 244 with 5 mismatches
Sequences output to file:
        from 62 to 182
                tcgagagccgtattgcgatatagcgcacatgcgttggacacagatgagca
                cacagtgacatgagagacacagatatagagacagatagacgatagacaga
                cagcatatatagacagatagc
        sequence trimmed from 5' end:
                TCTAGAACCGGTGACGTCTCCCATGGTGAAGCTTGGATCCACGATATCCT
                GCAGGAATTCC
        sequence trimmed from 3' end:
                TCGAGACCGTACGTGCGCGCGAATGCATCCAGATCTTCCCTCTAGTCAAG
                GCCTTAAGTGAGTCGTATTACGGA

        from 62 to 227
                tcgagagccgtattgcgatatagcgcacatgcgttggacacagatgagca
                cacagtgacatgagagacacagatatagagacagatagacgatagacaga
                cagcatatatagacagatagcTCGAGACCGTACGTGCGCGCGAATGCATC
```

```
                CAGATCTTCCCTCTAG
        sequence trimmed from 5' end:
                TCTAGAACCGGTGACGTCTCCCATGGTGAAGCTTGGATCCACGATATCCT
                GCAGGAATTCC
        sequence trimmed from 3' end:
                TCAAGGCCTTAAGTGAGTCGTATTACGGA
```

4. Running *vectorstrip* against a sequence containing Ns:

```
% vectorstrip pTYB1_N.seq
Strips out DNA between a pair of vector sequences
Are your vector sequences in a file? [Y]:
Name of vectorfile: vectors
Max allowed % mismatch [10]: 30
Show only the best hits (minimise mismatches)? [Y]:
Output file [ptyb1.vectorstrip]: stdout
Output sequence [ptyb1.fasta]:

Sequence: pTYB1.seq      Vector: pTYB1
5' sequence matches:
        From 40 to 58 with 2 mismatches
3' sequence matches:
        From 180 to 196 with 2 mismatches
Sequences output to file:
        from 59 to 179
                tcnagagccgtatngcgatatngcgcacatgcgntggacacagangagca
                cacagtnacatgagagncacagatntagagacagatngacgataganaga
                cagcatanatagacanatagc
        sequence trimmed from 5' end:
                CTTTAAGNAGGAGANATACANATGGCNAGCTCGCGANTCGACGGCGGNCG
                CGAATNCC
        sequence trimmed from 3' end:
                TCGNGGGCTCTTCCNGCTTTGCCANGGGTACCAANGTTTTAATGGCNGAT

Sequence: pTYB1.seq      Vector: pBS_KS+        No match

Sequence: pTYB1.seq      Vector: pLITMUS        No match
```

Mandatory qualifiers (bold if not always prompted):

[-sequence] (seqall)
: Sequence database USA.

[-[no]vectorfile] (boolean)
: Are your vector sequences in a file?

* **[-vectors]** *(infile)*
: Name of vectorfile.

-mismatch (integer)
: Max allowed percentage mismatch.

-[no]besthits (boolean)
: Show only the best hits (minimize mismatches)?

-linkera *(string)*
: 5' sequence.

-linkerb *(string)*
: 3' sequence.

[-outf] (outfile)
: Output filename.

[-outseq] (seqoutall)
: Output sequence(s) USA.

water

water uses the Smith-Waterman algorithm (modified for speed enhancements) to calculate the local alignment.

Here is a sample session with *water*:

```
% water tsw:hba_human tsw:hbb_human
Smith-Waterman local alignment.
Gap opening penalty [10.0]:
Gap extension penalty [0.5]:
Output alignment [hba_human.water]:
```

Mandatory qualifiers:

[-sequencea] (sequence)
: Sequence USA.

[-seqall] (seqall)
: Sequence database USA.

-gapopen (float)
: The gap open penalty is the score taken away when a gap is created. The best value depends on the choice of comparison matrix. The default value assumes you are using the EBLOSUM62 matrix for protein sequences, and the EDNAFULL matrix for nucleotide sequences.

-gapextend (float)
: The gap extension penalty is added to the standard gap penalty for each base or residue in the gap. This is how long gaps are penalized. You can usually expect a few long gaps rather than many short gaps, so the gap extension penalty should be lower than the gap penalty. An exception occurs when one or both sequences are single reads with possible sequencing errors, in which case you should expect many single base gaps. You can obtain this result by setting the gap open penalty to zero (or a very low value) and using the gap extension penalty to control gap scoring.

[-outfile] (align)
: Output alignment filename.

Optional qualifiers:

-datafile (matrix)
: This is the scoring matrix file used when comparing sequences. By default, it is the file *EBLOSUM62* (for proteins), or the file *EDNAFULL* (for nucleic sequences). These files are found in the *data* directory of the EMBOSS installation.

Advanced qualifiers:

-[no]similarity (boolean)
: Display percent identity and similarity.

whichdb

whichdb searches all available EMBOSS databases for sequences with a specified ID name or accession number.

Here is a sample session with *whichdb*:

```
% whichdb
Search all databases for an entry
ID or Accession number: hsfau
Output file [outfile.whichdb]:
```

Mandatory qualifiers (bold if not always prompted):

[-entry] (string)
ID or accession number.

-outfile (outfile)
Output filename.

Advanced qualifiers:

-get (boolean)
Retrieve sequences.

wobble

wobble plots the third position variability as an indicator of a potential coding region.

Here is a sample session with *wobble*. The example sequence is from *Pseudomonas aeruginosa*, which has a high G+C content and a very biased third codon position. If it can be G or C, it usually is:

```
% wobble
Wobble base plot
Input sequence: embl:paamir
Graph type [x11]:
Output file [paamir.wobble]:
```

Mandatory qualifiers:

[-sequence] (sequence)
Sequence USA.

-graph (xygraph)
Graph type.

-outf (outfile)
Output filename.

Optional qualifiers:

-window (integer)
Window size in codons.

Advanced qualifiers:

-bases (string)
Bases used.

wordcount

wordcount displays all the words of the specified length with the number of times they occur.

Here is a sample session with *wordcount*:

```
% wordcount embl:rnu68037 -wordsize=3
Counts words of a specified size in a DNA sequence
Output file [rnu68037.wordcount]:
```

Mandatory qualifiers:

[-sequence] (sequence)
: Sequence USA.

-wordsize (integer)
: Word size.

-outfile (outfile)
: Output filename.

wordmatch

wordmatch finds all exact matches of a given minimum size between 2 sequences displaying the start points in each sequence and the match length.

Here is a sample session with *wordmatch*:

```
% wordmatch tsw:hba_human tsw:hbb_human
Finds all exact matches of a given size between 2 sequences
Word size [4]:
Output alignment [hba_human.wordmatch]:
```

Mandatory qualifiers:

[-asequence] (sequence)
: Sequence USA.

[-bsequence] (sequence)
: Sequence USA.

-wordsize (integer)
: Word size.

[-outfile] (align)
: Output alignment filename.

Advanced qualifiers:

-afeatout (featout)
: File for output of normal tab-delimited GFF features.

-bfeatout (featout)
: File for output of normal tab-delimited GFF features.

wossname

wossname finds programs by keywords in their one-line documentation.

Here are some sample sessions with *wossname*:

Search for programs with "restrict" in their description:

```
% wossname restrict
```

Display a listing of programs in their groups:

```
% wossname -search ''
```

Display an alphabetic listing of all programs:

```
% wossname -search '' -alphabetic
```

Display only the groups that the programs can belong to:

```
% wossname -search '' -groups
```

Output html tags around the list of program groups:

```
% wossname '' -groups -html -prelink '#'
```

Output html tags around the list of programs:

```
% wossname '' -html -prelink 'http://www.sanger.ac.uk/Software/EMBOSS/Apps/'
-postlink '.shtml'
```

EMBOSS

Mandatory qualifiers:

[-search] (string)
: Enter a word or words here to perform a case-independent search of the one-line documentation provided by all EMBOSS programs. If no keyword is specified, all programs are listed.

Optional qualifiers (bold if not always prompted):

-explode (boolean)
: The groups EMBOSS applications belong to have two forms: exploded and non-exploded. The exploded group names are more numerous and often more vaguely phrased than the non-exploded ones. The exploded names are formed from definitions of the group names that start in the form `NAME1:NAME2` and are then expanded into many combinations of the names, such as: `NAME1, NAME2, NAME1 NAME2, NAME2 NAME1`. The non-expanded names are simply listed as: `NAME1 NAME2`. Using exploded group names finds many more programs that share at least some of the exploded names. More programs will be reported as sharing a similar function if you use use non-exploded names.

-outfile (outfile)
: If you enter the name of a file here, this program will write the program names and brief descriptions into that file.

-html (boolean)
: If you are sending the output to a file, this qualifier will format it for displaying as a table in a WWW document.

-prelink *(string)*
: If you are outputting to a file in HTML format, you can make the program names into hyperlinks by setting this qualifier to be the first half of the URL. If you want the program name to be a hyperlink to the URL *http://www.uk.embnet.org/Software/EMBOSS/Apps/programname.html*, enter `http://www.uk.embnet.org/Software/EMBOSS/Apps/` to set the first half of the URL before the program name.

-postlink *(string)*
: If you are outputting to a file in HTML format, you can make the program names into hyperlinks by setting this qualifier to be the second half of the URL. If you want the program name to be a hyperlink to the URL: *http://www.uk.embnet.org/*

Software/EMBOSS/Apps/programname.html, enter `.html` to set the second half of the URL after the program name.

-groups boolean
: If you use this option, only the group names will output to the file.

-alphabetic boolean
: If you use this option, you will get a single list of the program names and descriptions instead of the programs being listed in their functional groups.

Advanced qualifiers:

-[no]emboss (boolean)
: If you use this option, EMBOSS program documentation is searched. If this option is set to be false, only the EMBASSY programs are searched (if the `-embassy` option is true). EMBASSY programs are not strictly part of EMBOSS, but use the same code libraries and share the same look and feel. These programs are generally developed by people who want their programs to be outside of the GNU Public License scheme, i.e., they may want to charge a license fee.

-[no]embassy (boolean)
: If you use this option, EMBASSY program documentation is searched. If this option is set to be false, only the EMBOSS programs will be searched (if the `-emboss` option is true). EMBASSY programs are not strictly part of EMBOSS, but use the same code libraries and share the same look and feel, but are generally developed by people who want the programs to be outside of the GNU Public License scheme, i.e., they may want to charge a license fee.

-colon (boolean)
: The groups EMBOSS applications belong to may have up to two levels For example, the primary group ALIGNMENT has several sub-groups, or second-level groups: CONSENSUS, DIFFERENCES, DOT PLOTS, GLOBAL, LOCAL, and MULTIPLE. To aid programs that parse the output of *seealso* (and require the names of these subgroups), a colon is placed between the first and second level of the group name if this option is true.

-gui boolean
: This option is intended to help those who are designing Graphical User Interfaces (GUIs) to the EMBOSS applications. Some EMBOSS programs are inappropriate for running in a GUI, these include other menu programs and interactive editors. This option allows you to report only those programs that can be run from a GUI.

yank

yank is a simple utility that adds a specified sequence name to a list file.

Here is a sample session with *yank*, adding the regions making up the coding sequence of *embl:hsfau1* to the list *cds.list*:

```
% yank
Reads a range from a sequence, appends the full USA to a list file
Input sequence: em:hsfau1
     Begin at position [start]: 782
       End at position [end]: 856
        Reverse strand [N]:
Output file [hsfau1.yank]: cds.list
```

```
% yank
Reads a range from a sequence, appends the full USA to a list file
Input sequence: em:hsfau1
     Begin at position [start]: 951
       End at position [end]: 1095
        Reverse strand [N]:
Output file [hsfau1.yank]: cds.list

% yank
Reads a range from a sequence, appends the full USA to a list file
Input sequence: em:hsfau1
     Begin at position [start]: 1557
       End at position [end]: 1612
        Reverse strand [N]:
Output file [hsfau1.yank]: cds.list

% yank
Reads a range from a sequence, appends the full USA to a list file
Input sequence: em:hsfau1
     Begin at position [start]: 1787
       End at position [end]: 1912
        Reverse strand [N]:
Output file [hsfau1.yank]: cds.list
```

Mandatory qualifiers:

[-sequence] (sequence)
Sequence USA.

[-outfile] (outfile)
Output filename.

Advanced qualifiers:

-newfile (boolean)
Overwrite existing output file.

References

Rice, P., I. Longden, and A. Bleasby. 2000. EMBOSS: The European Molecular Biology Open Software Suite. *Trends in Genetics* 16 (6):276-277.

Main page
http://www.hgmp.mrc.ac.uk/Software/EMBOSS/

User documentation
http://www.hgmp.mrc.ac.uk/Software/EMBOSS/userdoc.html

Tutorial
http://www.hgmp.mrc.ac.uk/Software/EMBOSS/Doc/Tutorial/

Download
http://www.hgmp.mrc.ac.uk/Software/EMBOSS/download.html

Jemboss (Java API)
http://www.hgmp.mrc.ac.uk/Software/EMBOSS/Jemboss/

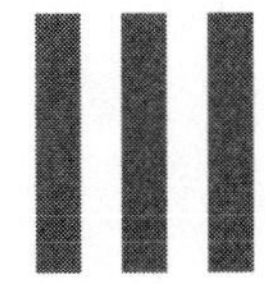

Appendixes

To complete this book, we are including a collection of very practical tables and resources anyone in bioinformatics can use.

The first two appendixes contain common tables for nucleotides, amino acids, and genetic codes. If you're like we are, you have versions of these thumbtacked near your workspace.

The references contain all references found at the end of individual chapters, plus our favorite books and some URLs for BioJava and BioPerl. One of us (Scott) has a pet peeve that references are never where you need them. So in our book, they're in two places.

In the final appendix, we mention our future plans by including a proposed contribution to the EMBOSS suite. We hope you like it as much as we enjoyed designing it.

Nucleotide and Amino Acid Tables

Trying to remember all of those single-letter codes for nucleotides and amino acids can be quite challenging, so we've compiled them into two useful tables. Nucleotides are listed in Table A-1; amino acids in Table A-2.

Nucleotide Codes

We have included a list of accepted nucleotide replacement combinations and their respective symbols. The symbols are commonly found in output data from sequencing instruments. Table A-1 shows the complete set of nucleotide codes.

Table A-1. Nucleotide codes

Symbol	Acceptable nucleotides	Name or type	Complement
G	G	Guanine	C
A	A	Adenine	T
T	T	Thymine	A
C	C	Cytosine	G
R	G or A	purine	Y
Y	T or C	pyrimidine	R
M	A or C	amino	K
K	G or T	keto	M
S	G or C		S
W	A or T		W
H	A or C or T		D
B	G or T or C		V
V	G or C or A		B
D	G or A or T		H
N	G or A or T or C	any nucleotide	N

Amino Acid Codes

Amino acids are commonly referred to as the "building blocks" of proteins. The characteristics of each amino acid are dependent on their side chain, and they can be divided into several classes. These classifications are denoted as either polar, nonpolar, acidic, or basic. Figure A-1 shows the structure of an amino acid.

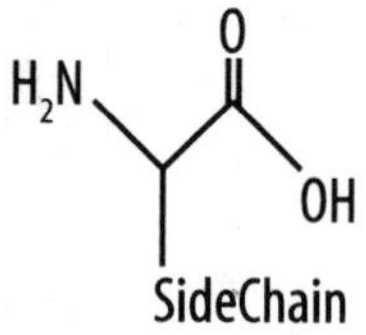

Figure A-1. Amino acid structure

Although trying to recall all of the amino acid names and their symbols may be a fun mental exercise, we've decided to include a table of amino acids. Table A-2 includes the names, three- and single-letter codes, side chains, and classification of each amino acid.

Table A-2. Amino acid codes

Name	Three-letter code	Single-letter code	Side chain	classification
Alanine	Ala	A	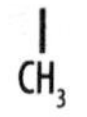 CH_3	nonpolar
Arginine	Arg	R	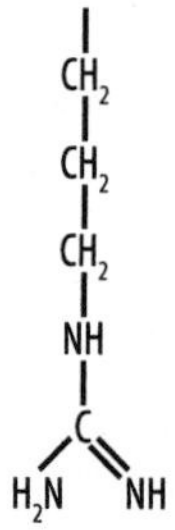 CH_2 – CH_2 – CH_2 – NH – C (H_2N, NH)	basic
Asparagine	Asn	N	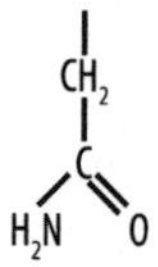 CH_2 – C (H_2N, O)	polar

Table A-2. Amino acid codes (continued)

Name	Three-letter code	Single-letter code	Side chain	classification
Aspartic Acid	Asp	D	$-CH_2-C(=O)OH$	acidic
Cysteine	Cys	C	$-CH_2-SH$	polar
Glutamine	Gln	Q	$-CH_2-CH_2-C(=O)NH_2$	polar
Glutamic Acid	Glu	E	$-CH_2-CH_2-C(=O)OH$	acidic
Glycine	Gly	G	$-H$	nonpolar
Histidine	His	H	$-CH_2-$ imidazole ring (C—N=CH—NH—HC=C)	basic

Table A-2. Amino acid codes (continued)

Name	Three-letter code	Single-letter code	Side chain	classification
Isoleucine	Ile	I	$H-C(-CH_3)-CH_2-CH_3$	nonpolar
Leucine	Leu	L	$CH_2-CH(CH_3)(H_3C)$	nonpolar
Lysine	Lys	K	$CH_2-CH_2-CH_2-CH_2-NH_2$	basic
Methionine	Met	M	$CH_2-CH_2-S-CH_3$	nonpolar
Phenylalanine	Phe	F	CH_2	nonpolar

Table A-2. Amino acid codes (continued)

Name	Three-letter code	Single-letter code	Side chain	classification
Proline	Pro	P	H, HN—C—COOH, H_2C CH_2, C, H_2	nonpolar
Selenocysteine	Sec	U	CH_2, Se	polar
Serine	Ser	S	H—C—OH, H	polar
Threonine	Thr	T	H—C—OH, CH_3	polar
Tryptophan	Trp	W	CH_2, C, HC, N, H	polar
Tyrosine	Tyr	Y	CH_2, OH	polar
Valine	Val	V	CH, H_3C CH_3	nonpolar

Table A-2. Amino acid codes (continued)

Name	Three-letter code	Single-letter code	Side chain	classification
Aspartic acid or Asparagine	Asx	B	See structures above.	-
Glutamic acid or Glutamine	Glx	Z	See structures above.	-
Unspecified Amino Acid	Xaa	X	-	-

Properties Summary

Although there are four main classifications for amino acids, other characteristics may also be used to describe them. This is best illustrated in the properties diagram in Figure A-2. In the diagram, *tiny* is used to describe very short side chains, while *small* is used to denote small side chains. The terms *aliphatic*, *aromatic* and *hydrophobic* commonly refer to the chemical composition of the side chain of an amino acid. These types of side chains are typically composed of only carbon and hydrogen atoms. The terms *charged*, *negative*, *positive*, and *polar* designate the electronic characteristics of a side chain. The diagram is adapted from Livingstone & Barton, *CABIOS*, **9**, 745-756, 1993 (*http://www.ncbi.nlm.nih.gov/htbin-post/Entrez/query?uid=8143162&form=6&db=m&Dopt=b PubMed*).

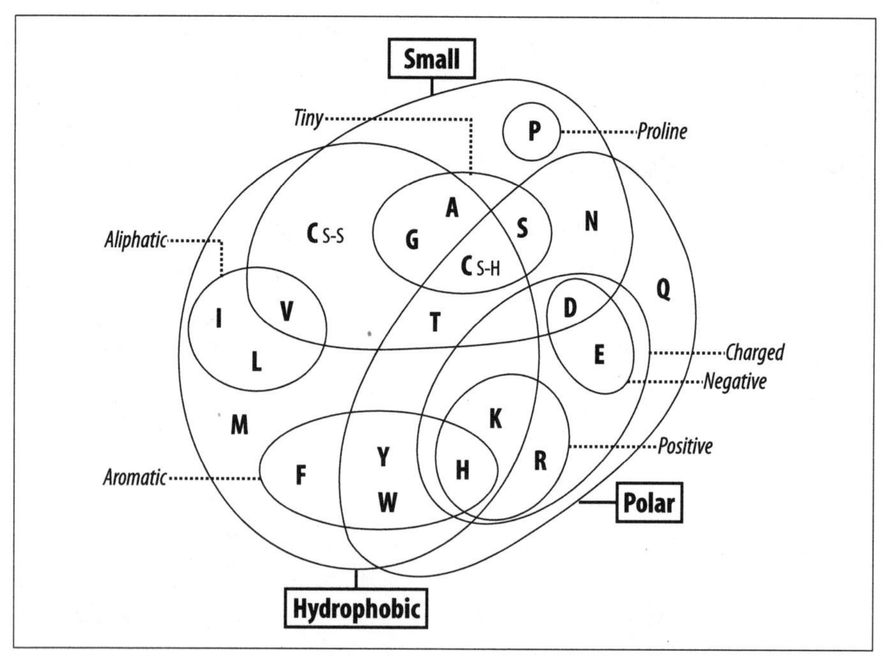

Figure A-2. Summary diagram of amino acid properties

References

Liébecq, C., ed. 1992. *Biochemical Nomenclature and Related Documents*, 2nd Edition. London: Portland Press.

Nucleic Acids

Nomenclature for Incompletely Specified Bases in Nucleic Acid Sequences (IUPAC IUBMB Joint Commission on Biochemical Nomenclature)

http://www.chem.qmw.ac.uk/iubmb/misc/naseq.html

Amino Acids

Nomenclature and Symbolism for Amino Acids and Peptides (IUPAC IUBMB Joint Commission on Biochemical Nomenclature)

http://www.chem.qmul.ac.uk/iupac/AminoAcid/

B

Genetic Codes

Amino acids are encoded by three-letter codons, but not all organisms have the same genetic code. Hence, we have included all of the common genetic codes.

The Standard Code

```
TTT F Phe      TCT S Ser      TAT Y Tyr      TGT C Cys
TTC F Phe      TCC S Ser      TAC Y Tyr      TGC C Cys
TTA L Leu      TCA S Ser      TAA * Ter      TGA * Ter
TTG L Leu i    TCG S Ser      TAG * Ter      TGG W Trp

CTT L Leu      CCT P Pro      CAT H His      CGT R Arg
CTC L Leu      CCC P Pro      CAC H His      CGC R Arg
CTA L Leu      CCA P Pro      CAA Q Gln      CGA R Arg
CTG L Leu i    CCG P Pro      CAG Q Gln      CGG R Arg

ATT I Ile      ACT T Thr      AAT N Asn      AGT S Ser
ATC I Ile      ACC T Thr      AAC N Asn      AGC S Ser
ATA I Ile      ACA T Thr      AAA K Lys      AGA R Arg
ATG M Met i    ACG T Thr      AAG K Lys      AGG R Arg

GTT V Val      GCT A Ala      GAT D Asp      GGT G Gly
GTC V Val      GCC A Ala      GAC D Asp      GGC G Gly
GTA V Val      GCA A Ala      GAA E Glu      GGA G Gly
GTG V Val      GCG A Ala      GAG E Glu      GGG G Gly
```

Vertebrate Mitochondrial Code

```
TTT F Phe     TCT S Ser     TAT Y Tyr     TGT C Cys
TTC F Phe     TCC S Ser     TAC Y Tyr     TGC C Cys
TTA L Leu     TCA S Ser     TAA * Ter     TGA W Trp
TTG L Leu     TCG S Ser     TAG * Ter     TGG W Trp

CTT L Leu     CCT P Pro     CAT H His     CGT R Arg
CTC L Leu     CCC P Pro     CAC H His     CGC R Arg
CTA L Leu     CCA P Pro     CAA Q Gln     CGA R Arg
CTG L Leu     CCG P Pro     CAG Q Gln     CGG R Arg

ATT I Ile i   ACT T Thr     AAT N Asn     AGT S Ser
ATC I Ile i   ACC T Thr     AAC N Asn     AGC S Ser
ATA M Met i   ACA T Thr     AAA K Lys     AGA * Ter
ATG M Met i   ACG T Thr     AAG K Lys     AGG * Ter

GTT V Val     GCT A Ala     GAT D Asp     GGT G Gly
GTC V Val     GCC A Ala     GAC D Asp     GGC G Gly
GTA V Val     GCA A Ala     GAA E Glu     GGA G Gly
GTG V Val i   GCG A Ala     GAG E Glu     GGG G Gly
```

The following table contains the differences in the Vertebrate Mitochondrial Code from the Standard Code.

Codon	Vertebrate Mitochondrial	Standard
AGA	* Ter	R Arg
AGG	* Ter	R Arg
AUA	M Met	I Ile
UGA	W Trp	* Ter

Yeast Mitochondrial Code

```
TTT F Phe     TCT S Ser     TAT Y Tyr     TGT C Cys
TTC F Phe     TCC S Ser     TAC Y Tyr     TGC C Cys
TTA L Leu     TCA S Ser     TAA * Ter     TGA W Trp
TTG L Leu     TCG S Ser     TAG * Ter     TGG W Trp

CTT T Thr     CCT P Pro     CAT H His     CGT R Arg
CTC T Thr     CCC P Pro     CAC H His     CGC R Arg
CTA T Thr     CCA P Pro     CAA Q Gln     CGA R Arg
CTG T Thr     CCG P Pro     CAG Q Gln     CGG R Arg
```

```
ATT I Ile      ACT T Thr    AAT N Asn    AGT S Ser
ATC I Ile      ACC T Thr    AAC N Asn    AGC S Ser
ATA M Met i    ACA T Thr    AAA K Lys    AGA R Arg
ATG M Met i    ACG T Thr    AAG K Lys    AGG R Arg

GTT V Val      GCT A Ala    GAT D Asp    GGT G Gly
GTC V Val      GCC A Ala    GAC D Asp    GGC G Gly
GTA V Val      GCA A Ala    GAA E Glu    GGA G Gly
GTG V Val      GCG A Ala    GAG E Glu    GGG G Gly
```

Comment: CGA and CGC are absent in this genetic code.

The following table contains the differences in the Yeast Mitochondrial Code from the Standard Code.

Codon	Yeast Mitochondrial	Standard
AUA	M Met	I Ile
CUU	T Thr	L Leu
CUC	T Thr	L Leu
CUA	T Thr	L Le3.u
CUG	T Thr	L Leu
UGA	W Trp	* Ter

Mold, Protozoan, and Coelenterate Mitochondrial Code and the Mycoplasma/Spiroplasma Code

```
TTT F Phe      TCT S Ser    TAT Y Tyr    TGT C Cys
TTC F Phe      TCC S Ser    TAC Y Tyr    TGC C Cys
TTA L Leu i    TCA S Ser    TAA * Ter    TGA W Trp
TTG L Leu i    TCG S Ser    TAG * Ter    TGG W Trp

CTT L Leu      CCT P Pro    CAT H His    CGT R Arg
CTC L Leu      CCC P Pro    CAC H His    CGC R Arg
CTA L Leu      CCA P Pro    CAA Q Gln    CGA R Arg
CTG L Leu i    CCG P Pro    CAG Q Gln    CGG R Arg

ATT I Ile i    ACT T Thr    AAT N Asn    AGT S Ser
ATC I Ile i    ACC T Thr    AAC N Asn    AGC S Ser
ATA I Ile i    ACA T Thr    AAA K Lys    AGA R Arg
ATG M Met i    ACG T Thr    AAG K Lys    AGG R Arg

GTT V Val      GCT A Ala    GAT D Asp    GGT G Gly
GTC V Val      GCC A Ala    GAC D Asp    GGC G Gly
GTA V Val      GCA A Ala    GAA E Glu    GGA G Gly
GTG V Val i    GCG A Ala    GAG E Glu    GGG G Gly
```

The following table contains the differences in the Mold, Protozoan, Coelenterate Mitochondrial, and *Mycoplasma/Spiroplasma* Code from the Standard Code.

Codon	Mold, Protozoan, and Coelenterate Mitochondrial Code and the Mycoplasma/Spiroplasma	Standard
UGA	W Trp	* Ter

Invertebrate Mitochondrial Code

```
TTT F Phe      TCT S Ser      TAT Y Tyr      TGT C Cys
TTC F Phe      TCC S Ser      TAC Y Tyr      TGC C Cys
TTA L Leu      TCA S Ser      TAA * Ter      TGA W Trp
TTG L Leu i    TCG S Ser      TAG * Ter      TGG W Trp

CTT L Leu      CCT P Pro      CAT H His      CGT R Arg
CTC L Leu      CCC P Pro      CAC H His      CGC R Arg
CTA L Leu      CCA P Pro      CAA Q Gln      CGA R Arg
CTG L Leu      CCG P Pro      CAG Q Gln      CGG R Arg

ATT I Ile i    ACT T Thr      AAT N Asn      AGT S Ser
ATC I Ile i    ACC T Thr      AAC N Asn      AGC S Ser
ATA M Met i    ACA T Thr      AAA K Lys      AGA S Ser
ATG M Met i    ACG T Thr      AAG K Lys      AGG S Ser

GTT V Val      GCT A Ala      GAT D Asp      GGT G Gly
GTC V Val      GCC A Ala      GAC D Asp      GGC G Gly
GTA V Val      GCA A Ala      GAA E Glu      GGA G Gly
GTG V Val i    GCG A Ala      GAG E Glu      GGG G Gly
```

Comment: The codon AGG is absent in *Drosophila*.

The following table contains the differences in the Invertebrate Mitochondrial Code from the Standard Code.

Codon	Invertebrate Mitochondrial	Standard
AGA	S Ser	R Arg
AGG	S Ser	R Arg
AUA	M Met	I Ile
UGA	W Trp	* Ter

Genetic Codes

Ciliate, Dasycladacean, and Hexamita Nuclear Code

```
TTT F Phe      TCT S Ser      TAT Y Tyr      TGT C Cys
TTC F Phe      TCC S Ser      TAC Y Tyr      TGC C Cys
TTA L Leu      TCA S Ser      TAA Q Gln      TGA * Ter
TTG L Leu      TCG S Ser      TAG Q Gln      TGG W Trp

CTT L Leu      CCT P Pro      CAT H His      CGT R Arg
CTC L Leu      CCC P Pro      CAC H His      CGC R Arg
CTA L Leu      CCA P Pro      CAA Q Gln      CGA R Arg
CTG L Leu      CCG P Pro      CAG Q Gln      CGG R Arg

ATT I Ile      ACT T Thr      AAT N Asn      AGT S Ser
ATC I Ile      ACC T Thr      AAC N Asn      AGC S Ser
ATA I Ile      ACA T Thr      AAA K Lys      AGA R Arg
ATG M Met i    ACG T Thr      AAG K Lys      AGG R Arg

GTT V Val      GCT A Ala      GAT D Asp      GGT G Gly
GTC V Val      GCC A Ala      GAC D Asp      GGC G Gly
GTA V Val      GCA A Ala      GAA E Glu      GGA G Gly
GTG V Val      GCG A Ala      GAG E Glu      GGG G Gly
```

The following table contains the differences in the Ciliate, Dasycladacean, and Hexamita Nuclear Code from the Standard Code.

Codon	Ciliate, Dasycladacean, and Hexamita Nuclear	Standard
UAA	Q Gln	* Ter
UAG	Q Gln	* Ter

Echinoderm and Flatworm Mitochondrial Code

```
TTT F Phe      TCT S Ser      TAT Y Tyr      TGT C Cys
TTC F Phe      TCC S Ser      TAC Y Tyr      TGC C Cys
TTA L Leu      TCA S Ser      TAA * Ter      TGA W Trp
TTG L Leu      TCG S Ser      TAG * Ter      TGG W Trp

CTT L Leu      CCT P Pro      CAT H His      CGT R Arg
CTC L Leu      CCC P Pro      CAC H His      CGC R Arg
CTA L Leu      CCA P Pro      CAA Q Gln      CGA R Arg
CTG L Leu      CCG P Pro      CAG Q Gln      CGG R Arg
```

```
ATT I Ile      ACT T Thr      AAT N Asn      AGT S Ser
ATC I Ile      ACC T Thr      AAC N Asn      AGC S Ser
ATA I Ile      ACA T Thr      AAA N Asn      AGA S Ser
ATG M Met i    ACG T Thr      AAG K Lys      AGG S Ser

GTT V Val      GCT A Ala      GAT D Asp      GGT G Gly
GTC V Val      GCC A Ala      GAC D Asp      GGC G Gly
GTA V Val      GCA A Ala      GAA E Glu      GGA G Gly
GTG V Val i    GCG A Ala      GAG E Glu      GGG G Gly
```

The following table contains the differences in the Echinoderm and Flatworm Mitochondrial Code from the Standard Code

Codon	Echinoderm and Flatworm Mitochondrial	Standard
AAA	N ASN	K Lys
AGA	S Ser	R Arg
AGG	S Ser	R Arg
UGA	W Trp	* Ter

Euplotid Nuclear Code


```
TTT F Phe      TCT S Ser      TAT Y Tyr      TGT C Cys
TTC F Phe      TCC S Ser      TAC Y Tyr      TGC C Cys
TTA L Leu      TCA S Ser      TAA * Ter      TGA C Cys
TTG L Leu      TCG S Ser      TAG * Ter      TGG W Trp

CTT L Leu      CCT P Pro      CAT H His      CGT R Arg
CTC L Leu      CCC P Pro      CAC H His      CGC R Arg
CTA L Leu      CCA P Pro      CAA Q Gln      CGA R Arg
CTG L Leu      CCG P Pro      CAG Q Gln      CGG R Arg

ATT I Ile      ACT T Thr      AAT N Asn      AGT S Ser
ATC I Ile      ACC T Thr      AAC N Asn      AGC S Ser
ATA I Ile      ACA T Thr      AAA K Lys      AGA R Arg
ATG M Met i    ACG T Thr      AAG K Lys      AGG R Arg

GTT V Val      GCT A Ala      GAT D Asp      GGT G Gly
GTC V Val      GCC A Ala      GAC D Asp      GGC G Gly
GTA V Val      GCA A Ala      GAA E Glu      GGA G Gly
GTG V Val      GCG A Ala      GAG E Glu      GGG G Gly
```

The following table contains the differences in the Euplotid Nuclear Code from the Standard Code.

Codon	Euplotid Nuclear	Standard
UGA	C Cys	* Ter

Bacterial and Plant Plastid Code

```
TTT F Phe      TCT S Ser      TAT Y Tyr      TGT C Cys
TTC F Phe      TCC S Ser      TAC Y Tyr      TGC C Cys
TTA L Leu      TCA S Ser      TAA * Ter      TGA * Ter
TTG L Leu i    TCG S Ser      TAG * Ter      TGG W Trp

CTT L Leu      CCT P Pro      CAT H His      CGT R Arg
CTC L Leu      CCC P Pro      CAC H His      CGC R Arg
CTA L Leu      CCA P Pro      CAA Q Gln      CGA R Arg
CTG L Leu i    CCG P Pro      CAG Q Gln      CGG R Arg

ATT I Ile i    ACT T Thr      AAT N Asn      AGT S Ser
ATC I Ile i    ACC T Thr      AAC N Asn      AGC S Ser
ATA I Ile i    ACA T Thr      AAA K Lys      AGA R Arg
ATG M Met i    ACG T Thr      AAG K Lys      AGG R Arg

GTT V Val      GCT A Ala      GAT D Asp      GGT G Gly
GTC V Val      GCC A Ala      GAC D Asp      GGC G Gly
GTA V Val      GCA A Ala      GAA E Glu      GGA G Gly
GTG V Val i    GCG A Ala      GAG E Glu      GGG G Gly
```

Alternative Yeast Nuclear Code

```
TTT F Phe      TCT S Ser      TAT Y Tyr      TGT C Cys
TTC F Phe      TCC S Ser      TAC Y Tyr      TGC C Cys
TTA L Leu      TCA S Ser      TAA * Ter      TGA * Ter
TTG L Leu      TCG S Ser      TAG * Ter      TGG W Trp

CTT L Leu      CCT P Pro      CAT H His      CGT R Arg
CTC L Leu      CCC P Pro      CAC H His      CGC R Arg
CTA L Leu      CCA P Pro      CAA Q Gln      CGA R Arg
CTG S Ser i    CCG P Pro      CAG Q Gln      CGG R Arg

ATT I Ile      ACT T Thr      AAT N Asn      AGT S Ser
ATC I Ile      ACC T Thr      AAC N Asn      AGC S Ser
ATA I Ile      ACA T Thr      AAA K Lys      AGA R Arg
ATG M Met i    ACG T Thr      AAG K Lys      AGG R Arg
```

```
GTT V Val     GCT A Ala     GAT D Asp     GGT G Gly
GTC V Val     GCC A Ala     GAC D Asp     GGC G Gly
GTA V Val     GCA A Ala     GAA E Glu     GGA G Gly
GTG V Val     GCG A Ala     GAG E Glu     GGG G Gly
```

THe following table contains the differences in the Alternative Yeast Nuclear Code from the Standard Code.

Codon	Alternative Yeast Nuclear	Standard
CUG	S Ser	L Leu

Ascidian Mitochondrial Code

```
TTT F Phe     TCT S Ser     TAT Y Tyr     TGT C Cys
TTC F Phe     TCC S Ser     TAC Y Tyr     TGC C Cys
TTA L Leu     TCA S Ser     TAA * Ter     TGA W Trp
TTG L Leu     TCG S Ser     TAG * Ter     TGG W Trp

CTT L Leu     CCT P Pro     CAT H His     CGT R Arg
CTC L Leu     CCC P Pro     CAC H His     CGC R Arg
CTA L Leu     CCA P Pro     CAA Q Gln     CGA R Arg
CTG L Leu     CCG P Pro     CAG Q Gln     CGG R Arg

ATT I Ile     ACT T Thr     AAT N Asn     AGT S Ser
ATC I Ile     ACC T Thr     AAC N Asn     AGC S Ser
ATA M Met     ACA T Thr     AAA K Lys     AGA G Gly
ATG M Met i   ACG T Thr     AAG K Lys     AGG G Gly

GTT V Val     GCT A Ala     GAT D Asp     GGT G Gly
GTC V Val     GCC A Ala     GAC D Asp     GGC G Gly
GTA V Val     GCA A Ala     GAA E Glu     GGA G Gly
GTG V Val     GCG A Ala     GAG E Glu     GGG G Gly
```

The following table contains the differences in the Ascidian Mitochondrial Code from the Standard Code.

Codon	Ascidian Mitochondrial	Standard
AGA	G Gly	R Arg
AGG	G Gly	R Arg
AUA	M Met	I Ile
UGA	W Trp	* Ter

Genetic Codes

Alternative Flatworm Mitochondrial Code

```
TTT F Phe     TCT S Ser     TAT Y Tyr     TGT C Cys
TTC F Phe     TCC S Ser     TAC Y Tyr     TGC C Cys
TTA L Leu     TCA S Ser     TAA Y Tyr     TGA W Trp
TTG L Leu     TCG S Ser     TAG * Ter     TGG W Trp

CTT L Leu     CCT P Pro     CAT H His     CGT R Arg
CTC L Leu     CCC P Pro     CAC H His     CGC R Arg
CTA L Leu     CCA P Pro     CAA Q Gln     CGA R Arg
CTG L Leu     CCG P Pro     CAG Q Gln     CGG R Arg

ATT I Ile     ACT T Thr     AAT N Asn     AGT S Ser
ATC I Ile     ACC T Thr     AAC N Asn     AGC S Ser
ATA I Ile     ACA T Thr     AAA N Asn     AGA S Ser
ATG M Met i   ACG T Thr     AAG K Lys     AGG S Ser

GTT V Val     GCT A Ala     GAT D Asp     GGT G Gly
GTC V Val     GCC A Ala     GAC D Asp     GGC G Gly
GTA V Val     GCA A Ala     GAA E Glu     GGA G Gly
GTG V Val     GCG A Ala     GAG E Glu     GGG G Gly
```

The following table contains the differences in the Alternative Flatworm Mitochondrial Code from the Standard Code.

Codon	Alternative Flatworm Mitochondrial	Standard
AAA	N Asn	K Lys
AGA	S Ser	R Arg
AGG	S Ser	R Arg
UAA	Y Tyr	* Ter
UGA	W Trp	* Ter

Blepharisma Nuclear Code

```
TTT F Phe     TCT S Ser     TAT Y Tyr     TGT C Cys
TTC F Phe     TCC S Ser     TAC Y Tyr     TGC C Cys
TTA L Leu     TCA S Ser     TAA * Ter     TGA * Ter
TTG L Leu     TCG S Ser     TAG Q Gln     TGG W Trp

CTT L Leu     CCT P Pro     CAT H His     CGT R Arg
CTC L Leu     CCC P Pro     CAC H His     CGC R Arg
CTA L Leu     CCA P Pro     CAA Q Gln     CGA R Arg
CTG L Leu     CCG P Pro     CAG Q Gln     CGG R Arg
```

```
ATT I Ile       ACT T Thr       AAT N Asn       AGT S Ser
ATC I Ile       ACC T Thr       AAC N Asn       AGC S Ser
ATA I Ile       ACA T Thr       AAA K Lys       AGA R Arg
ATG M Met i     ACG T Thr       AAG K Lys       AGG R Arg

GTT V Val       GCT A Ala       GAT D Asp       GGT G Gly
GTC V Val       GCC A Ala       GAC D Asp       GGC G Gly
GTA V Val       GCA A Ala       GAA E Glu       GGA G Gly
GTG V Val       GCG A Ala       GAG E Glu       GGG G Gly
```

The following table contains the differences in the *Blepharisma* Nuclear Code from the Standard Code.

Codon	Blepharisma Nuclear	Standard
UAG	Q Gln	* Ter

Chlorophycean Mitochondrial Code

```
TTT F Phe       TCT S Ser       TAT Y Tyr       TGT C Cys
TTC F Phe       TCC S Ser       TAC Y Tyr       TGC C Cys
TTA L Leu       TCA S Ser       TAA * Ter       TGA * Ter
TTG L Leu       TCG S Ser       TAG L Leu       TGG W Trp

CTT L Leu       CCT P Pro       CAT H His       CGT R Arg
CTC L Leu       CCC P Pro       CAC H His       CGC R Arg
CTA L Leu       CCA P Pro       CAA Q Gln       CGA R Arg
CTG L Leu       CCG P Pro       CAG Q Gln       CGG R Arg

ATT I Ile       ACT T Thr       AAT N Asn       AGT S Ser
ATC I Ile       ACC T Thr       AAC N Asn       AGC S Ser
ATA I Ile       ACA T Thr       AAA K Lys       AGA R Arg
ATG M Met i     ACG T Thr       AAG K Lys       AGG R Arg

GTT V Val       GCT A Ala       GAT D Asp       GGT G Gly
GTC V Val       GCC A Ala       GAC D Asp       GGC G Gly
GTA V Val       GCA A Ala       GAA E Glu       GGA G Gly
GTG V Val       GCG A Ala       GAG E Glu       GGG G Gly
```

The following table contains the differences in the Chlorophycean Mitochondrial Code from the Standard Code.

Codon	Chlorophycean Mitochondrial	Standard
UAG	L Leu	* Ter

Genetic Codes

Trematode Mitochondrial Code

```
TTT F Phe      TCT S Ser      TAT Y Tyr      TGT C Cys
TTC F Phe      TCC S Ser      TAC Y Tyr      TGC C Cys
TTA L Leu      TCA S Ser      TAA * Ter      TGA W Trp
TTG L Leu      TCG S Ser      TAG * Ter      TGG W Trp

CTT L Leu      CCT P Pro      CAT H His      CGT R Arg
CTC L Leu      CCC P Pro      CAC H His      CGC R Arg
CTA L Leu      CCA P Pro      CAA Q Gln      CGA R Arg
CTG L Leu      CCG P Pro      CAG Q Gln      CGG R Arg

ATT I Ile      ACT T Thr      AAT N Asn      AGT S Ser
ATC I Ile      ACC T Thr      AAC N Asn      AGC S Ser
ATA M Met      ACA T Thr      AAA N Asn      AGA S Ser
ATG M Met i    ACG T Thr      AAG K Lys      AGG S Ser

GTT V Val      GCT A Ala      GAT D Asp      GGT G Gly
GTC V Val      GCC A Ala      GAC D Asp      GGC G Gly
GTA V Val      GCA A Ala      GAA E Glu      GGA G Gly
GTG V Val i    GCG A Ala      GAG E Glu      GGG G Gly
```

The following table contains the differences in the Trematode Mitochondrial Code from the Standard Code.

Codon	Trematode Mitochondrial	Standard
AAA	N Asn	K Lys
AGA	S Ser	R Arg
AGG	S Ser	R Arg
AUA	M Met	I Ile
UGA	W Trp	* Ter

Scenedesmus Obliquus Mitochondrial Code

```
TTT F Phe      TCT S Ser      TAT Y Tyr      TGT C Cys
TTC F Phe      TCC S Ser      TAC Y Tyr      TGC C Cys
TTA L Leu      TCA * Ter      TAA * Ter      TGA * Ter
TTG L Leu      TCG S Ser      TAG L Leu      TGG W Trp

CTT L Leu      CCT P Pro      CAT H His      CGT R Arg
CTC L Leu      CCC P Pro      CAC H His      CGC R Arg
CTA L Leu      CCA P Pro      CAA Q Gln      CGA R Arg
CTG L Leu      CCG P Pro      CAG Q Gln      CGG R Arg
```

```
ATT I Ile        ACT T Thr        AAT N Asn        AGT S Ser
ATC I Ile        ACC T Thr        AAC N Asn        AGC S Ser
ATA I Ile        ACA T Thr        AAA K Lys        AGA R Arg
ATG M Met i      ACG T Thr        AAG K Lys        AGG R Arg

GTT V Val        GCT A Ala        GAT D Asp        GGT G Gly
GTC V Val        GCC A Ala        GAC D Asp        GGC G Gly
GTA V Val        GCA A Ala        GAA E Glu        GGA G Gly
GTG V Val        GCG A Ala        GAG E Glu        GGG G Gly
```

The following table contains the differences in the *Scenedesmus Obliquus* Mitochondrial Code from the Standard Code.

Codon	Scenedesmus Obliquus Mitochondrial	Standard
UAG	L Leu	* Ter
UCA	* Ter	S Ser

Thraustochytrium Mitochondrial Code

```
TTT F Phe        TCT S Ser        TAT Y Tyr        TGT C Cys
TTC F Phe        TCC S Ser        TAC Y Tyr        TGC C Cys
TTA * Ter        TCA S Ser        TAA * Ter        TGA * Ter
TTG L Leu        TCG S Ser        TAG * Ter        TGG W Trp

CTT L Leu        CCT P Pro        CAT H His        CGT R Arg
CTC L Leu        CCC P Pro        CAC H His        CGC R Arg
CTA L Leu        CCA P Pro        CAA Q Gln        CGA R Arg
CTG L Leu        CCG P Pro        CAG Q Gln        CGG R Arg

ATT I Ile i      ACT T Thr        AAT N Asn        AGT S Ser
ATC I Ile        ACC T Thr        AAC N Asn        AGC S Ser
ATA I Ile        ACA T Thr        AAA K Lys        AGA R Arg
ATG M Met i      ACG T Thr        AAG K Lys        AGG R Arg

GTT V Val        GCT A Ala        GAT D Asp        GGT G Gly
GTC V Val        GCC A Ala        GAC D Asp        GGC G Gly
GTA V Val        GCA A Ala        GAA E Glu        GGA G Gly
GTG V Val i      GCG A Ala        GAG E Glu        GGG G Gly
```


The following table contains the differences in the *Thraustochytrium* Mitochondrial Code from the Standard Code.

Codon	Thraustochytrium Mitochondrial	Standard
UUA	* Ter	L Leu

References

Compiled by Andrzej (Anjay) Elzanowski and Jim Ostell. National Center for Biotechnology Information (NCBI), Bethesda, MD. Last update of the Genetic Codes: October 05, 2000.

Genetic codes online
http://www.ncbi.nlm.nih.gov/htbin-post/Taxonomy/wprintgc?mode=t

C
Resources

The following references are those we used to pull the material together for this Nutshell. The web site and book lists are by no means comprehensive, but should give you a good start. Of course, now that you've got this Nutshell, you probably won't need to go digging nearly as often!

Web Sites

The organization of web sites follows that of the book: data, tools, and useful tables. Each site's main page is listed, as are URLs for documentation (release notes, README files, etc.) and downloads.

Data Formats

DDBJ

Main site
: *http://www.ddbj.nig.ac.jp/*

Release notes
: *http://www.ddbj.nig.ac.jp/ddbjnew/ddbj_relnote.html*

Download
: *ftp://ftp.ddbj.nig.ac.jp/database/ddbj/*

EMBL

Main page
: *http://www.ebi.ac.uk/embl/index.html*

Release notes
: *http://www.ebi.ac.uk/embl/Documentation/Release_notes/current/relnotes.html*

User manual
http://www.ebi.ac.uk/embl/Documentation/User_manual/usrman.html

Download
ftp://ftp.ebi.ac.uk/pub/databases/embl/

GenBank

GenBank overview
http://www.ncbi.nlm.nih.gov/Genbank/GenbankOverview.html

Release notes
ftp://ftp.ncbi.nih.gov/genbank/gbrel.txt

Download
ftp://ftp.ncbi.nih.gov/genbank/

DDBJ/EMBL/GenBank Feature Table
http://www.ncbi.nlm.nih.gov/projects/collab/FT/index.html

NCBI Sequence Identifier Syntax
ftp://ftp.ncbi.nih.gov/blast/db/README

Non-redundant database
ftp://ftp.ncbi.nih.gov/blast/db/README

Pfam

Main page
http://pfam.wustl.edu/

Release notes
ftp://ftp.genetics.wustl.edu/pub/Pfam/relnotes.txt

Help pages
http://pfam.wustl.edu/help.shtml

Download
ftp://ftp.genetics.wustl.edu/pub/Pfam/

PROSITE

Main page
http://us.expasy.org/prosite/

Release notes
http://us.expasy.org/prosite/psrelnot.html

User manual
http://us.expasy.org/prosite/prosuser.html

Download
ftp://us.expasy.org/databases/prosite/

SWISS-PROT

Main page
http://us.expasy.org/sprot/

Release notes
http://us.expasy.org/sprot/relnotes/

User manual

http://us.expasy.org/sprot/userman.html

Download

ftp://us.expasy.org/databases/swiss-prot

Tools

BLAST

Main page

http://www.ncbi.nlm.nih.gov/BLAST/

Information guide

http://www.ncbi.nlm.nih.gov/Education/BLASTinfo/information3.html

Download

ftp://ftp.ncbi.nih.gov/blast/executables/

BLAT

Main page

http://genome.ucsc.edu/cgi-bin/hgBlat?command=start

User guide

http://genome.ucsc.edu/goldenPath/help/hgTracksHelp.html

Download

http://www.soe.ucsc.edu/~kent/exe/

Clustal

Main page

http://www.ebi.ac.uk/clustalw/

User help

http://www.ebi.ac.uk/clustalw/clustalw_frame.html

Download

ftp://ftp.ebi.ac.uk/pub/software/unix/clustalw/
ftp://ftp.ebi.ac.uk/pub/software/dos/clustalw/

HMMER

Main page

http://hmmer.wustl.edu/

README

ftp://ftp.genetics.wustl.edu/pub/eddy/hmmer/CURRENT/00README

Download

ftp://ftp.genetics.wustl.edu/pub/eddy/hmmer/2.2g/hmmer-2.2g.tar.gz

EMBOSS

Main page

http://www.hgmp.mrc.ac.uk/Software/EMBOSS/

User documentation
: *http://www.hgmp.mrc.ac.uk/Software/EMBOSS/userdoc.html*

Tutorial
: *http://www.hgmp.mrc.ac.uk/Software/EMBOSS/Doc/Tutorial/*

Download
: *http://www.hgmp.mrc.ac.uk/Software/EMBOSS/download.html*

Jemboss (Java API)
: *http://www.hgmp.mrc.ac.uk/Software/EMBOSS/Jemboss/*

Common Tables

Nucleic Acids

Nomenclature for Incompletely Specified Bases in Nucleic Acid Sequences (IUPAC IUBMB Joint Commission on Biochemical Nomenclature)

http://www.chem.qmw.ac.uk/iubmb/misc/naseq.html

Genetic Codes

http://www.ncbi.nlm.nih.gov/htbin-post/Taxonomy/wprintgc?mode=t

Amino Acids

Nomenclature and Symbolism for Amino Acids and Peptides (IUPAC IUBMB Joint Commission on Biochemical Nomenclature)

http://www.chem.qmul.ac.uk/iupac/AminoAcid/

Miscellaneous

BioPerl

Main page
: *http://bioperl.org/*

User documentation
: *http://bioperl.org/Core/Latest/modules.html*

Mailing lists
: *http://bioperl.org/MailList.shtml*

Download
: *http://bioperl.org/Core/Latest/index.shtml*

BioJava

Main page
: *http://biojava.org/*

User documentation
: *http://biojava.org/docs/started.html*
http://biojava.org/docs/api/

Mailing lists
: *http://biojava.org/mailman/listinfo/biojava-l*
http://biojava.org/mailman/listinfo/biojava-dev

Download
: *http://biojava.org/download/*

The WWW Virtual Library: Model Organisms

http://www.ceolas.org/VL/mo/

Books

The following books are the ones we use most.

Baxevanis, Andreas D. (Editor) and B. F. Francis Ouellette. 2001. *Bioinformatics: A Practical Guide to the Analysis of Genes and Proteins, Second Edition*. New York: Wiley-Interscience.

Branden, Carl-Ivar, and John Tooze. 1999. *Introduction to Protein Structure, 2nd Edition*. New York: Garland Publishing.

Durbin, Richard, Sean Eddy, Anders Krogh, and Graeme Mitchison. 1998. *Biological Sequence Analysis: Probabilistic Models of Proteins and Nucleic Acids*. Cambridge: Cambridge University Press.

Flanagan, David. 2002. *Java in a Nutshell, 4th Edition*. Sebastopol: O'Reilly.

Friedl, Jeffrey E. F. 2002. *Mastering Regular Expressions, 2nd Edition*. Sebastopol: O'Reilly.

Gibas, Cynthia, and Per Jambeck. 2001. *Developing Bioinformatics Computer Skills*. Sebastopol: O'Reilly.

Gonick, Larry, and Mark Wheelis. 1991. *The Cartoon Guide to Genetics*. New York: Harper Perennial.

Gusfield, Dan. 1997. *Algorithms on Strings, Trees, and Sequences: Computer Science and Computational Biology*. Cambridge: Cambridge University Press.

Liébecq, C., ed. 1992. *Biochemical Nomenclature and Related Documents, 2nd Edition*. London: Portland Press.

Mount, David W. 2001. *Bioinformatics: Sequence and Genome Analysis*. Woodbury, New York: Cold Spring Harbor Laboratory Press.

Robbins, Arnold. 1999. *UNIX in a Nutshell: System V Edition, 3rd Edition*. Sebastopol: O'Reilly.

Tisdall, James. 2001. *Beginning Perl for Bioinformatics*. Sebastopol: O'Reilly.

Wall, Larry, Jon Orwant, and Tom Christiansen. 2000. *Programming Perl, 3rd Edition*. Sebastopol: O'Reilly.

Watson, James D., Alan M. Weiner, and Nancy H. Hopkins. 2001. *Molecular Biology of the Gene, 4th Edition*. New York: Addison Wesley.

Watson, James D., Michael Gilman, Jan Witkowski, Mark Zoller, and Gilman Witkowski. 1992. *Recombinant DNA*. New York: W H Freeman & Co.

Journal Articles

Altschul, S.F., W. Gish, W. Miller, E. W. Myers, and D. J. Lipman. 1990. Basic local alignment search tool. *J. Mol. Biol.* 215:403–410.

Altschul, S.F., T. L. Madden, A. A. Schäffer, J. Zhang, Z. Zhang, W. Miller, and D. J. Lipman. 1997. Gapped BLAST and PSI-BLAST: a new generation of protein database search programs. *Nucleic Acids Research* 25:3389–3402.

Bailey, Timothy L., and Charles Elkan. 1994 Fitting a mixture model by expectation maximization to discover motifs in biopolymers. *Proceedings of the Second International Conference on Intelligent Systems for Molecular Biology* 28–36. Menlo Park: AAAI Press.

Bailey, Timothy L., and Michael Gribskov. 1998. Combining evidence using p-values: application to sequence homology searches. *Bioinformatics* 14:48–54.

Bairoch, A., and R. Apweiler. 2000. The SWISS-PROT protein sequence database and its supplement TrEMBL in 2000. *Nucleic Acids Research* 28:45-48.

Bateman, A., E. Birney, L. Cerruti, R. Durbin, L. Etwiller, S. R. Eddy, S. Griffiths-Jones, K. L. Howe, M. Marshall, and E. L. L. Sonnhammer. 2002. The Pfam Protein Families Database. *Nucleic Acids Research* 30 (1):275–280.

Benson, D.A., I. Karsch-Mizrachi, D. J. Lipman, J. Ostell, B. A. Rapp, and D. L. Wheeler. 2002. GenBank. *Nucleic Acids Research* 30 (1):17–20.

Falquet L, Pagni M, Bucher P, Hulo N, Sigrist CJA, Hofmann K, Bairoch A. 2002. The PROSITE database, its status in 2002. *Nucleic Acids Research.* Jan 1;30(1):235–8.

Gilbert, D. G. 1999. Readseq version 2, an improved biosequence conversion tool, written in the Java language. *Bionet.Software* (August).

Gish, W., and D. J. States. 1993. Identification of protein coding regions by database similarity search. *Nature Genet.* 3:266–272.

Higgins, D., J. Thompson, T. Gibson, J. D. Thompson, D. G. Higgins, T. J. Gibson. 1994. CLUSTAL W: improving the sensitivity of progressivemultiple sequence alignment through sequence weighting,position-specific gap penalties and weight matrix choice. *Nucleic Acids Research* 22:4673–4680.

International Human Genome Sequencing Consortium. 2001. Initial sequencing and analysis of the human genome. *Nature* 409:860–921.

Kent, W. James. 2002. BLAT - The BLAST-Like Alignment Tool. *Genome Research* 12 (4):656–664.

Pearson, W.R., and D. J. Lipman. 1988. Improved Tools for Biological Sequence Analysis. *Proceedings of the National Academy of Sciences* 85:2444–2448.

Rice, P., I. Longden, and A. Bleasby. 2000. EMBOSS: The European Molecular Biology Open Software Suite. *Trends in Genetics* 16 (6):276–277.

Sigrist CJA, Cerutti L, Hulo N, Gattiker A, Falquet L, Pagni M, Bairoch A, Bucher P. 2002. PROSITE: a documented database using patterns and profiles as motif descriptors. *Brief Bioinform.* 3(3):265–74.

Stoesser, G., W. Baker, A. van den Broek, E. Camon, M. Garcia-Pastor, C. Kanz, T. Kulikova, R. Leinonen, Q. Lin, V. Lombard, R. Lopez, N. Redaschi, P. Stoehr, M. A. Tuli, K. Tzouvara, and R. Vaughan. 2002. The EMBL Nucleotide Sequence Database. *Nucleic Acids Research* 30 (1):21–26.

Tateno, Y., T. Imanishi, S. Miyazaki, K. Fukami-Kobayashi, N. Saitou, H. Sugawara, and T. Gojobori. 2002. DNA Data Bank of Japan (DDBJ) for genome scale research in life science. *Nucleic Acids Research* 30 (1):27–30.

Venter, J. Craig, et al. 2001. The Sequence of the Human Genome. *Science* 291:1304-1351.

D

Future Plans

Here's a program we're working on. When it's finished we plan to donate it to the EMBOSS package. It may take us a while, so be patient.

crystalball

crystalball answers every question that you want to ask about a target sequence. It's used by heads of drug discovery, CIOs, and CFOs at all of the major pharmaceutical companies.

Here's a sample *crystalball* session:

```
% crystalball mySeq -rdtime -rdcost -profit -outfile stdout
Answers every drug discovery question you have about this sequence.
Drug discovery development time: 10-15 years
Drug discovery development cost: $800 million*
Profit: Not nearly as much as you'd like!
```

Mandatory qualifiers

[-sequence] (sequence)
: Sequence USA.

[-outfile] (outfile)
: Output filename.

Optional qualifiers

-competition (boolean)
: Who else is working with this target?

-rdtime (boolean)
: Total research and development time to bring a drug for this target to market.

* Time and cost estimate from "Tufts Center for the Study of Drug Development Pegs Cost of a New Prescription Medicine at $802 Million." Tufts University press release (November 30, 2001).

-rdcost (boolean)
Total cost of our research and development effort.

-animalstudies (boolean)
What will we learn from the animal studies?

-clinicaltrials (boolean)
Detail all of the surprises we'll get from the clinical trials.

-fdaproblems (boolean)
List all of the issues the FDA will raise with our paperwork.

-fdatime (boolean)
How long will the FDA take to render a decision?

-profit (boolean)
How much will we make after the drug gets to market?

Advanced qualifiers

-everythingelse (boolean)
Tell us everything else we'd really like to know now rather than later.

In case you haven't guessed, *crystalball* is a completely fictitious routine. EMBOSS doesn't contain this functionality. No one else has it either, but lots of people wish for just such a tool!

Future Plans

Index

We'd like to hear your suggestions for improving our indexes. Send email to *index@oreilly.com*.

C

D

F

G

H

I

J

K

L

M

N

O

P

T

U

V

W

Y

About the Authors

Scott Markel is a Principal Software Architect at LION bioscience Inc., where he is responsible for providing architectural direction in the development of software for the life sciences, including the use and development of standards. He is a co-chair of the Life Sciences Research Domain Task Force of the Object Management Group and also chairs the LSR's Architecture and Roadmap Working Group. Prior to working at LION, Scott worked at NetGenics, Johnson & Johnson Pharmaceutical Research & Development, and Sarnoff Corporation. He has a Ph.D. in mathematics from the University of Wisconsin-Madison. When Scott's not working or writing, he enjoys spending time with his wife and kids, reading European history books, and just enjoying life in sunny San Diego.

Darryl León is a Principal Scientific Architect at LION bioscience Inc., where he is responsible for providing scientific direction in the development of software for the life sciences. Prior to working at LION, Darryl worked at NetGenics, DoubleTwist, and Genset. He has taught at California Polytechnic State University, San Luis Obispo, and currently teaches a bioinformatics class at U.C. Santa Cruz Extension and U.C. San Diego Extension. He is also a member of the Bioinformatics Advisory Committee at U.C. San Diego Extension. Darryl has a Ph.D. in biochemistry from the University of California-San Diego and did his postdoctoral research at the University of California-Santa Cruz.

Colophon

Our look is the result of reader comments, our own experimentation, and feedback from distribution channels. Distinctive covers complement our distinctive approach to technical topics, breathing personality and life into potentially dry subjects.

The animal on the cover of *Sequence Analysis in a Nutshell: A Guide to Common Tools and Databases* is a liger. Much like sequence analysis, which is a cross between computer science and biology, a liger is the result of a cross between a lion and a tigress. Tigons, which are the offspring of a tiger and a lioness, have also been bred.

In a classic display of "hybrid vigor," a liger may be 10 to 12 feet in length and weigh upwards of 800 to 1,000 pounds, making it significantly larger than either of its parents. Because lions and tigers are genetically very similar, their offspring show a fascinating blend of the features and habits of both species. Depending on which subspecies of lion and tiger are bred together, and how their genes combine, a liger may look more tigerish or more lionish. Ligers are both striped and spotted, the spots being inherited from the lion. A male liger may grow a leonine mane or the facial ruff of a tiger. Female ligers exhibit both the leonine need for social interaction and the tiger-like need for solitude. Both sexes roar like lions and "chuff" like tigers, and most ligers also inherit their tiger parents' love of water. Like other interspecies hybrids, ligers are usually sterile.

Ligers do not occur in the wild, but are the result of captive breeding. Natural breeding is not impossible, just extremely unlikely because the habitats of tigers and lions have little to no overlap, and the solitary tiger would most likely avoid interacting with a pride of lions.

Philip Dangler was the production editor and copyeditor for *Sequence Analysis in a Nutshell: A Guide to Common Tools and Databases*. Emily Quill, Linley Dolby, and Claire Cloutier provided quality control. Judy Hoer provided production assistance. Nancy Crumpton wrote the index.

Ellie Volckhausen designed the cover of this book, based on a series design by Edie Freedman. The cover image is an original illustration created by Lorrie LeJeune. Emma Colby produced the cover layout with QuarkXPress 4.1 using Adobe's ITC Garamond font.

David Futato designed the interior layout. This book was converted by Joe Wizda and Mike Sierra to FrameMaker 5.5.6 with a format conversion tool created by Erik Ray, Jason McIntosh, Neil Walls, and Mike Sierra that uses Perl and XML technologies. The text font is Linotype Birka; the heading font is Adobe Myriad Condensed; and the code font is LucasFont's TheSans Mono Condensed. The illustrations that appear in the book were produced by Robert Romano and Jessamyn Read using Macromedia FreeHand 9 and Adobe Photoshop 6. This colophon was written by Lorrie LeJeune.

Other Titles Available from O'Reilly

Bioinformatics

Beginning Perl for Bioinformatics

By James Tisdall
1st Edition October 2001
384 pages, 0-596-00080-4

This book shows biologists with little or no programming experience how to use Perl, the ideal language for biological data analysis. Each chapter focuses on solving a particular problem or class of problems, so you'll finish the book with a solid understanding of Perl basics, a collection of programs for such tasks as parsing BLAST and GenBank, and the skills to tackle more advanced bioinformatics programming.

Developing Bioinformatics Computer Skills

By Cynthia Gibas & Per Jambeck
1st Edition April 2001
446 pages, 1-56592-664-1

Developing Bioinformatics Computer Skills will help biologists, researchers, and students develop a structured approach to biological data and the computer tools they'll need to analyze it. The book covers the Unix file system, building tools and databases for bioinformatics, computational approaches to biological problems, an introduction to Perl for bioinformatics, data mining, data visualization, and tips for tailoring data analysis software to individual research needs.

Learning the UNIX Operating System, 5th Edition

By Jerry Peek, Grace Todino & John Strang
5th Edition November 2001
174 pages, ISBN 0-596-00261-0

Learning the UNIX Operating System is the most effective introduction to Unix in print. The fifth edition covers Internet usage for email, file transfers, and web browsing. It's perfect for those who are just starting with Unix or Linux, as well as anyone who encounters a Unix system on the Internet. Complete with a quick-reference card to pull out and keep handy, it's an ideal primer for Mac and PC users of the Internet who need to know a little bit about Unix on the systems they visit.

UNIX in a Nutshell: System V Edition, 3rd Edition

By Arnold Robbins
3rd Edition September 1999
616 pages, ISBN 1-56592-427-4

The bestselling, most informative Unix reference book is now more complete and up-to-date. Not a scaled-down quick reference of common commands, *UNIX in a Nutshell* is a complete reference containing all commands and options, with descriptions and examples that put the commands in context. For all but the thorniest Unix problems, this one reference should be all you need. Covers System V Release 4 and Solaris 7.

Learning Perl, 3rd Edition

By Randal Schwartz & Tom Phoenix
3rd Edition July 2001
330 pages, ISBN 0-596-00132-0

Learning Perl is the quintessential tutorial for the Perl programming language. The third edition has not only been updated to Perl Version 5.6, but has also been rewritten from the ground up to reflect the needs of programmers learning Perl today. Other books may teach you to program in Perl, but this book will turn you into a Perl programmer.

Web Database Applications with PHP & MySQL

By Hugh E. Williams & David Lane
1st Edition March 2002
582 pages, ISBN 0-596-00041-3

This book offers both theoretical and practical guidance for creating web database applications. The detailed information on designing relational databases and the web application architectures that interact with them will be especially useful to readers who have worked with or built database-backed web sites before. The book implements a sample web application using PHP and MySQL on the Apache platform.

How to stay in touch with O'Reilly

1. Visit our award-winning web site

http://www.oreilly.com/

★ "Top 100 Sites on the Web"—PC Magazine
★ CIO Magazine's Web Business 50 Awards

Our web site contains a library of comprehensive product information (including book excerpts and tables of contents), downloadable software, background articles, interviews with technology leaders, links to relevant sites, book cover art, and more. File us in your bookmarks or favorites!

2. Join our email mailing lists

Sign up to get email announcements of new books and conferences, special offers, and O'Reilly Network technology newsletters at:

http://elists.oreilly.com

It's easy to customize your free elists subscription so you'll get exactly the O'Reilly news you want.

3. Get examples from our books

To find example files for a book, go to:

http://www.oreilly.com/catalog

select the book, and follow the "Examples" link.

4. Work with us

Check out our web site for current employment opportunites:

http://jobs.oreilly.com/

5. Register your book

Register your book at:

http://register.oreilly.com

6. Contact us

O'Reilly & Associates, Inc.
1005 Gravenstein Hwy North
Sebastopol, CA 95472 USA
TEL: 707-827-7000 or 800-998-9938
(6am to 5pm PST)
FAX: 707-829-0104

order@oreilly.com
For answers to problems regarding your order or our products. To place a book order online visit:

http://www.oreilly.com/order_new/

catalog@oreilly.com
To request a copy of our latest catalog.

booktech@oreilly.com
For book content technical questions or corrections.

corporate@oreilly.com
For educational, library, government, and corporate sales.

proposals@oreilly.com
To submit new book proposals to our editors and product managers.

international@oreilly.com
For information about our international distributors or translation queries. For a list of our distributors outside of North America check out:

http://international.oreilly.com/distributors.html

adoption@oreilly.com
For information about academic use of O'Reilly books, visit:

http://academic.oreilly.com

To order: *800-998-9938* • *order@oreilly.com* • *www.oreilly.com*
Online editions of most O'Reilly titles are available by subscription at *safari.oreilly.com*
Also available at most retail and online bookstores.